Food Science and Technology

FOOD HYDROCOLLOID EDIBLE FILMS AND COATINGS

FOOD SCIENCE AND TECHNOLOGY

Food Science and Technology: New Research
Lorenzo V. Greco and Marco N. Bruno (Editors)
2008. 978-1-60456-715-1

Food Science and Technology: New Research
Lorenzo V. Greco and Marco N. Bruno (Editors)
2008. 978-1-61668-106-7

The Price of Food
Meredith N. Fisher (Editor)
2009. 978-1-60692-440-2

Food Processing and Engineering Topics
Maria Elena Sosa-Morales and Jorge F. Velez-Ruiz (Editors)
2009. 978-1-60741-788-0

Traditional Chinese Foods: Production and Research Progress
Li Zaigui and Tan Hongzhuo (Authors)
2009. 978-1-60692-902-5

Food Science Research and Technology
Isaak Hülsen and Egon Ohnesorge (Editors)
2009. 978-1-60741-848-1

Food, Diet and Health: Past, Present and Future Tendencies
Raquel Pinho Ferreira Guine (Editor)
2009. 978-1-60876-012-1

Food Processing and Engineering Topics
Maria Elena Sosa-Morales and Jorge F. Velez-Ruiz (Editors)
2009. 978-1-91998-422-8

Traditional Chinese Foods: Production and Research Progress
Li Zaigui and Tan Hongzhuo (Authors)
2009. 978-1-61668-277-4

Food Science Research and Technology
Isaak Hülsen and Egon Ohnesorge (Editors)
2009. 978-1-61668-853-0

Food Safety, Fresh Produce and FDA Oversight
Russell P. Barton (Editor)
2010. 978-1-60692-788-5

Industrial Fermentation: Food Processes, Nutrient Sources and Production Strategies
Jürgen Krause and Oswald Fleischer (Editors)
2010. 978-1-60876-550-8

Food Hydrocolloids: Characteristics, Properties and Structures
Clarence S. Hollingworth (Editor)
2010. 978-1-60876-222-4

Sweet Potato: Post Harvest Aspects in Food, Feed and Industry
Ramesh C. Ray and K.I. Tomlins (Editors)
2010. 978-1-60876-343-6

Microbial Interaction in Fermented Beverages
Marta E. Farías, Oscar A. Sosa, Lucía M. Mendoza and Pedro A. Aredes Fernández (Authors)
2010. 978-1-60876-785-4

The Olive: Its Processing and Waste Management
José S. Torrecilla (Author)
2010. 978-1-60876-719-9

Liquid Chromatography for the Determination of Mycotoxins in Foods
R. Romero-González, J. L. Martínez Vidal and A. Garrido Frenich (Authors)
2010. 978-1-60876-882-0

Applications of Natural Products in Food
Supayang Piyawan Voravuthikunchai and Beatrice Olawumi T. Ifesan (Editors)
2010. 978-1-60876-998-8

Advances in Food Science and Technology
A. K. Haghi (Editor)
2010. 978-1-61668-415-0

Extraction of Metabolities from Plant Tissues
A. Mohdaly and I. Smetanska (Authors)
2010. 978-1-61668-252-1

Food Hydrocolloid Edible Films and Coatings
O. Skurtys, C. Acevedo, F. Pedreschi, J. Enronoe, F. Osorio and J.M. Aguilera (Authors)
2010. 978-1-61668-269-9

Fermented Milk Products
R. Ahmed Abdelrahman, A.M Adel and I. Smetanska (Authors)
2010. 978-1-61668-299-6

Applications of Natural Products in Food
Supayang Piyawan Voravuthikunchai and Beatrice Olawumi T. Ifesan (Editors)
2010. 978-1-61668-370-2

Extraction of Metabolities from Plant Tissues
A. Mohdaly and I. Smetanska (Authors)
2010. 978-1-61668-616-1

Food Hydrocolloid Edible Films and Coatings
O. Skurtys, C. Acevedo, F. Pedreschi, J. Enronoe, F. Osorio and J.M. Aguilera (Authors)
2010. 978-1-61668-740-3

Fermented Milk Products
R. Ahmed Abdelrahman, A.M Adel and I. Smetanska (Authors)
2010. 978-1-61668-741-0

Food Science and Technology

FOOD HYDROCOLLOID EDIBLE FILMS AND COATINGS

O. SKURTYS

C. ACEVEDO

F. PEDRESCHI

J. ENRONOE

F. OSORIO

AND

J.M. AGUILERA

Nova Science Publishers, Inc.

New York

For permission to use material from this book please contact us:
Telephone 631-231-7269; Fax 631-231-8175
Web Site: http://www.novapublishers.com

NOTICE TO THE READER

LIBRARY OF CONGRESS CATALOGING-IN-PUBLICATION DATA

Available upon Request
ISBN: 978-1-61668-269-9

Published by Nova Science Publishers, Inc. ✢ New York

CONTENTS

PREFACE

In this chapter book, we present and discuss some aspects of hydrocolloid edible films and coating. The film-forming materials is presented and the physical properties (*e.g.* thickness, mechanical, optical and thermodynamic properties, water and gases barrier) of the film and coating are examined in detail. The main techniques used to form edible coatings as spray systems or immersion procedures are discussed as well as the solvent casting and the extrusion processes for the edible films. The main disadvantage of these techniques is the lost of quality of the edible coatings and films since there is no control over the shape, size and size distribution of the dispersed elements (*e.g.* additives, ingredients, etc.) in the support structure matrix is poor. Thus, microfluidic applications are presented and discussed to solve some of these problems and to improve film qualities.

Chapter 1

1. INTRODUCTION

An edible film is defined as a thin layer, which can be consumed, coated on a food or placed as barrier between the food and the surrounding environment. For the past 10 years, research on edible films and coatings in foods is driven by food engineers due to the high demand of consumers for longer shelf–life and better quality of fresh foods as well as of environmentally friendly packagings [1-4]. The most familiar example of edible packaging is sausage meat in casing that is not removed for cooking and eating. Such films can mechanically protect foods, prevent the contamination from microorganisms, prevent quality loss of foods due to mass transfer (*e.g.* moisture, gases, flavours, etc.). Indeed, edible films and coatings can be used as a vehicle for incorporating natural or chemical antimicrobial agents, antioxidants, enzymes or functional ingredients such as probiotics, minerals and vitamins [5-7]. The edible films are classified into three categories taking into account the nature of their components: hydrocolloids (containing proteins, polysaccharides or alginates), lipids (constituted by fatty acids, acylglycerols or waxes) and composites (made by combining substances from the two categories) [8].

Thus, the objective of this work is to present and discuss only some aspects of hydrocolloid edible films and coatings. In particular, the film-forming materials is presented and the physical properties (*e.g.* thickness, mechanical, optical and thermodynamic properties, water and gases barrier) of the film and coating are examined. The main techniques used to form edible coatings as spray systems or immersion procedures are detailed as well as the solvent casting and the extrusion processes for the edible films. The main disadvantage of these techniques is the lost of quality of the edible coatings and films since there is no control over the shape, size and size distribution of the dispersed elements (e.g. additives, ingredients, etc.) in the support structure matrix is poor. An other disadvantage of these techniques is that the thickness of the films is generally no constant and no controlled. Finally, microfluidic applications are presented and discussed to solve some of these problems and to improve film qualities.

Chapter 2

2. Film-Forming Materials

2.1. Hydrocolloidal Materials

Hydrocolloids are hydrophilic polymers, of vegetable, animal, microbial or synthetic origin, that generally contain many hydroxyl groups and may be polyelectrolytes (for example alginate, carrageenan, carboxymethylcellulose, gum arabic, pectin and xanthan gum). Nowadays, they are widely used as film-forming solution to perform and control the texture, flavour, and shelf-life of foods [9]. Hydrocolloidal materials, *i.e.* proteins and polysaccharides, used extensively for the formation of edible films and coatings are presented in Table 1. All hydrocolloids used are fully or partially soluble in water and are used principally to increase the viscosity of the continuous phase (aqueous phase) *i.e* as gelling agent or thickener [1, 8, 10-11]. They can also be used as emulsifier since their stabilizing effect on emulsions derives from an increase in viscosity of the aqueous phase of the edible film. The kinetic motion of the oil droplets is reduced, resulting in a lower rate of flocculation and coalescence in the film.

2.1.1. Polysaccharides-Based Films and Coating

Agar (E406)

Agar is a hydrophilic colloid consisting of a mixture of agarose and agaropectin that have the ability to form reversible gels simply by cooling a hot aqueous solution. Agar gel melts on heating and resets on cooling. Typical gel temperatures of agar for 1.5% solution are in the range 35-45 °C [12]. Because of its ability to form very hard gels at very low concentrations and the simplicity of the extraction process [13], agar has been used extensively as a gelling agent in the food industry. However, in despite its biodegradability and its enormous gelling power, agar has been few used as edible film due to a poor aging. Indeed, both photodegradation and fluctuations in ambient temperature and humidity alter agar crystallinity, leading to formation of micro-fractures and polymer embrittlement [14-15]. However, it was reported that agar-based film displays a better moisture barrier properties than cassava starch film [16]. The influence of agar on the structure and the functional properties of emulsified edible films has been recently studied by Phan The *et al.* [17]. Gelled agar chains can stabilize film-forming emulsion to create a macronetwork. This macronetwork entraps flattened lipid particles improving barrier performance by increasing tortuosity. The mechanical properties of agar-hydrogenated vegetable oil emulsified films are comparable with some protein and Low-density polyethylene (LDPE) films.

Alginate (E400-E404)

Alginate is an indigestible biomaterial produced by brown seaweeds (Phaeophyceae, mainly Laminaria) therefore it may also be viewed as a source of dietary fibre. Alginate has a potential to form biopolymer film or coating component because of its unique colloidal properties, which include thickening, stabilizing, suspending, film forming, gel producing, and emulsion stabilizing [18-19]. An attractive feature of alginate solutions is the gelling capacity in presence of Calcium. Edible films prepared from alginate form strong films and exhibit poor water resistance because of their hydrophilic nature [10, 20-21]. The water permeability and mechanical attributes can be considered as moderate compared to synthetic films [22]. Alginate edible-films are appropriated to load additives and antibacterial compounds. Good results are obtained applying probiotics [23] and oil compunds such as garlic oil [24] and oregano oil [25]. A mixture of starch and alginate to form edible film has been studied by Wu *et al.* [26] achieving to improve the mechanical properties of film.

Carrageenans (E407)

Carrageenans are water-soluble polymers with a linear chain of partially sulphated galactans, which present high potentiality as film-forming material. These sulphated polysaccharides are extracted from the cell walls of various red seaweeds (Rhodophyceae). Different seaweeds produce different carrageenans. The positions and numbers of sulfate ester groups are important because they are, together with the anhydrogalactose bridge, responsible for carrageenan functionality and determine a classification in three major types: κ, ι and λ. κ- and ι-carrageenans contain the 3,6-anhydro units and are used as gelling agents because of their property to produce thermoreversible gels on cooling below the critical temperature. λ-Carrageenan, with only sulfated galactose groups and no anhydrogalactose bridge, is a thickening polymer [27]. The use of carrageenan as edible films and coatings already covers various fields of the food industry such as application on fresh and frozen meat, poultry and fish to prevent superficial dehydration [28], sausage-casings [29], dry solids foods, oily foods [30], etc. Carrageenan film formation includes this gelation mechanism during moderate drying, leading to a three-dimensional network formed by polysaccharide double-helices and to a solid film after solvent evaporation [31-32]. Recently, carrageenan films were also found to be less opaque than those made of starch [33].

Cellulose Derivatives

Cellulose derivatives are polysaccharides composed of linear chains of β (1–4) glucosidic units with methyl, hydroxypropyl or carboxyl substituents. Only four cellulose derivative forms are used for edible coatings or films: Hydroxypropyl cellulose (E463; HPC), hydroxypropyl methylcellulose (E464; HPMC), Carboxymethylcellulose (E466; CMC) or Methyl cellulose (E461; MC). Cellulose derivatives exhibit thermo-gelation therefore when suspensions are heated they form a gel whereas they returns to originally consistency when cooled [34]. The films cast from aqueous solutions of MC, HPMC, HPC and CMC tend to have moderate strength, are resistant to oils and fats, and are flexible, transparent, flavourless, colourless, tasteless, water-soluble and moderate barriers to oxygen. MC is the most resistant to water and it is the lowest hydrophilic cellulose derivatives [10]. However, cellulose derivative films are poor water vapour barriers because of the inherent hydrophilic nature of polysaccharides and they possess poor mechanical properties [35-36]. A way to improve the moisture barrier would be the incorporation of hydrophobic compounds, such as fatty acids, into the cellulose ether matrix to develop a composite film [37]. As other example, Debeaufort *et al.* [38], developed bilayers by adding a mixture of lipids (paraffin oil, paraffin wax, or a mixture of hydrogenated palm oil and triolein) onto a methylcellulose layer.

Edible coatings made of CMC, MC, HPC, and HPMC have been applied to some fruits and vegetables for providing barriers to oxygen, oil, or moisture transfer [7, 39].

Chitosan

Chitosan (β-(1, 4)-2-amino-2-deoxy-D-glucopyranose), which is mainly made from crustacean shells, is the second most abundant natural and non-toxic polymer in nature after cellulose [40-42]. Chitosan shows antifungal and antibacterial properties, which are believed to be originated from its polycationic nature [41, 43-44]. However, a major drawback of chitosan is its poor solubility in neutral solutions. The required degree of deacetylation to obtain a soluble product must be 80–85% or higher [45]. Chitosan products are highly viscous, resembling natural gums [46]. Chitosan can form transparent films to enhance the quality and extend the storage life of food products [33, 47-50]. Pure chitosan films are generally cohesive, compact and the film surface has a smooth contour without pores or cracks [51]. Chitosan films such as many polysaccharide based films, tend to exhibit fat and oil resistance and selective permeability to gases but lack resistance to water transmission [52-53]. Chitosan-based edible films and coatings allowed to extend and preserve shelf life of fresh strawberries [54-55], of slices mango fruit [56]. Finally, there is an available review on chitosan films [57].

Gums

Gums in edible-forming preparation are used a for their texturizing capabilities. All gums are polysaccharides composed of sugars other than glucose. Gums are differentiate in three groups [9]: exudate gums (gum Arabic; mesquite gum), the extractive gums (come from endosperm of some legume seeds or extracted from the wood: guar gum) and the microbial fermentation gums (xanthan gum). In edible-forming preparations, guar gum (E 412) is used as a water binder, stabilizer and viscosity builder. Gum arabic (E 414), owing to its solubility in hot or cold water, is the least viscous of the hydrocolloid gums. Xanthan gum (E 415) is readily dispersed in water, hence high consistency is obtained rapidly in both hot and cold systems. A blend of guar gum, gum arabic and xanthan gum provided uniform coatings with good cling and improved adhesion in wet batters [58-60]. The coating based from mesquite gum preparation and applied to Persian limes provided a lowest physiological weight loss, a best dark shade green colour retention and an unaltered physicochemical parameters to Persian limes [61]. The mesquite gum forms films with excellent water vapor barrier properties when small amounts of lipids are added in their formulation [62]. Film-forming polymers based on gum arabic, showed to be an inhibitor of after-cooking darkening of potatoes [63].

Pectin

Pectin (E440) is a heterogeneous grouping of acidic structural polysaccharides, found in fruit and vegetables and mainly prepared from citrus peel and apple pomace. This complex anionic polysaccharide is composed of β-1,4-linked d-galacturonic acid residues, wherein the uronic acid carboxyls are either fully (HMP, high methoxy pectin) or partially (LMP, low methoxy pectin) methyl esterified. Commercial extraction causes extensive degradation of the neutral sugar-containing side chains. With Chitosan, HMP or LMP forms excellent films. Indeed, the cationic nature of chitosan offers the possibility to take advantage of the electrostatic interactions with anionic polyelectrolytes, such as pectin [64-65]. The effect of antimicrobial plant essential oils such as oregano, lemon grass, or cinnamon oil at different concentrations, on physical properties of apple puree-pectin based edible coatings was reported by Rojas-Grau *et al.* [66]. It is also interesting to point out the application of LMP based edible coatings as a pre-treatment in osmotic dehydration for obtaining a better dehydration efficiency [67-68]. Finally recently, an investigations demonstrated the prevention of crumb ageing of dietetic sucrose-free sponge cake when a pectin-containing edible film was used. This sponge cake had better preserved freshness, especially up to the fifth day of storage [69].

Starch

Starch is the major carbohydrate reserve in plant tubers and seed endosperm where it is found as granules, each typically containing several million amylopectin molecules accompanied by a much larger number of smaller amylose molecules [70]. Amylose is responsible for the film-forming capacity of starch [71]. The largest source of starch is corn (maize) with other commonly used sources being wheat, potato, tapioca and rice. Genetic modification of starch crops has recently led to the development of starches with improved and targeted functionality [72]. Starch is used to produce biodegradable films to partially or entirely replace plastic polymer [73]. The films are transparent or translucent, flavourless, tasteless and colourless [74]. However, starch film application are limited by poor mechanical strength and its efficient barrier against low polarity compound [10]. Many research reported that film forming conditions have an effect on crystallinity of the starch films and, therefore, their properties [75]. Films of high-amylose corn starch or potato starch was more stable during aging, lost little of their elongation and had not or a slight increasing in tensile strength [76]. Films from cassava starch had good flexibility and low water permeability, indicating potential application as edible films [77]. Plasticizer is generally required for starch-based edible films to overcome film brittleness. The most commonly used plasticizers for starch films are glycerol and sorbitol [78]. Recently, Hu *et al.* [79] have developed starch films from oxidized potato starch (OPS) with glycerol as a plasticizer at different contents. The OPS films were transparent and flexible with interesting mechanical properties.

2.1.2. Proteins-Based Films and Coatings

Animal (milk and whey proteins) and plant (such as soy and zein proteins) protein films exhibit better oxygen barrier, carbon dioxide barrier and mechanical properties than polysaccharide films. However, similar to polysaccharide films, due to their hydrophillic nature, protein films generally possess poor water vapor barrier property [80]. Studies have shown that the concentration, composition, size, and shape of plasticizers all affect the properties of protein films [81-85]. Proteins that are insoluble in water, such as corn zein and wheat gluten, produce insoluble coatings, whereas proteins that are soluble in water produce coatings of varying solubility, depending on the protein and the conditions of coating formation and treatment. Most of the protein in soybeans is insoluble in water but soluble in dilute neutral salt solutions therefore soy protein are not considered as a hydrocolloid.

Gelatin

Gelatin is prepared by the thermal denaturation of collagen, isolated from animal skin, bones and fish skins [86]. The physical properties of gelatins are related not only to the molecular weight distribution but also to the amino acid composition. Gelatin is readily soluble in water at temperatures above 40ºC, forming a viscous solution of random-coiled linear polypeptide chains. On cooling a gelatin solution around 20ºC, collagen like helices are formed, albeit not very long ones and including only part of the material: this is form a gel [70]. Mammalian gelatins commonly have better physical properties and thermostability than most fish gelatins [87], and this has been related mainly to their higher amino acid content [86]. The use of mammalian gelatin in the elaboration of edible film or coatings was very well studied until the sixties, which resulted in many patents, mainly in the pharmaceutical area. Consequently, these materials characteristics are not easily available in the literature [88]. The properties and film forming ability of gelatins are directly related to the molecular weight, i.e., the higher the average molecular weight, the better the quality of the film [89]. The molecular weight distribution depends mainly on the degree of collagen cross-linking and the extraction procedure [90]. However, in the years 2000, the gelatin films forming principally with fish gelatin have returned to the attention of researcher [91-94]. Films from tuna skin gelatin plasticized with glycerol presented lower water vapor permeability (WVP) compared to values reported for pigskin gelatin [95].

Whey Proteins

Whey protein isolate produces totally water-soluble coatings but heat denatured solutions of whey protein isolate produce coatings in which the protein is insoluble [96]. Industrially produced whey protein concentrate (WPC) have a protein content between 25 and 80% whereas the whey protein isolate (WPI) have protein content above 90% [97]. Whey proteins have been the subject of intense investigation over the past decade or so [98]. Whey proteins can produce transparent, flexible, colourless and flavourless films, with a poor moisture barrier [99-100]. Moreover protein-based films possess a good aroma barrier [101], and low oxygen permeability [102]. Whey protein films produced without addition of any plasticizers are very brittle therefore the addition of plasticizers provides flexibility to the films but also increases their water vapour permeability. Increased concentration of plasticizers in edible whey protein films decreased tensile strength and Young's modulus, but increased elongation [103]. The water vapour permeability can also be improved by the incorporation of hydrophobic materials such as lipids [104]. However, the incorporation of lipid materials into edible film formulations into films could adversely affect the sensory characteristics of films. The incorporation of antimicrobial agents into edible coating formulations is needed to obtain stronger inhibitory effect against microbial growth (See paragraph 2.3). Optimization of edible whey protein films containing sorbitol, beeswax and potassium sorbate concentrations for mechanical and optical properties was reported by Ozdemir and Floros [105]. Recently WPI and mesquite gum (MG) are showed totally compatible to form solutions and films. Indeed, the incorporation of different amounts of MG on the WPI allowed to obtain composite films with improved flexibility without increasing plasticizer content and consequently without affecting negatively others characteristics such as water vapour permeability [106].

Table 1. Hydrocolloidal materials that have been studied extensively for the formation of edible films and coatings in foods

Polysaccharide	Film Forming Material		Principal Function	References
	Agar (E406)		Gelling agent	[14-17]
	Alginate (E400-404)		Gelling agent	[10, 19-24]
	Carrageenan (E407)		Gelling agent	[26-29, 31-32]
	Cellulose Derivative	Carboxymethyl cellulose (E466)	Thickner	[7, 10, 36, 38-39, 107-112]
		Hydroxypropyl cellulose (E463)	Thickner and emulsifier	
		Hydroxypropyl methyl (E464)	Thickner	
		Methyl cellulose (E461)	Thickner, emulsifier and gelling agent	
	Chitosan		Gelling agent	[33, 47-50, 53-56, 113]
			Antimicrobials	
	Gum	Arabic gum (E414)	Emulsifier	[58-63, 114]
		Guar gum (E412)	Thickner	
Protein		Xanthan gum (E415)	Thickner	
	Pectin (E440)		Gelling agent	[67-69, 115]
	Starches		Thickner and gelling agent	[76-79, 116-117]
	Gelatin	Bovine gelatin	Gelling agent	[88, 92, 94]
		Fish gelatin	Gelling agent	[92, 95, 118-119]
		Pig gelatin	Gelling agent	[88, 92-93]
	Whey Protein			[96, 102-103, 105-106]

2.2. PLASTICIZERS

Edible films and coatings need to have good elasticity and flexibility, a low brittleness, a high toughness and to prevent cracking during handling and storage [120]. Therefore, plasticizers of low molecular weight (non volatile) are typically added to hydrocolloid film forming solutions to modify the flexibility of edible films. Plasticizers with characteristics such as small size, high polarity, more polar groups per molecule, and greater distance between polar groups within a molecule generally impart greater plasticizing effects on a polymeric system. Indeed, they act by increasing the free volume or in other word by decreasing intermolecular attractions between adjacent polymeric chains by reducing hydrogen bonding between polymers chains [121]. Generally, plasticizers are required for polysaccharides (or proteins) based edible films. Their amount added into hydrocolloid film-forming preparations varies between 10% and 60% by weight of the hydrocolloid. The most commonly used plasticizers are: polyols (propylene glycol [122], glycerol [121], sorbitol [123], polyethylene glycol [124]), oligosaccharides (sucrose [125]) and water. In fact, water is a ubiquitous natural diluent, which plasticizes and/or antiplasticizes some films depending on the amount sorbed onto the films matrix [126]. Thus, film moisture content, as affected by the relative humidity of the surrounding environment, largely affects film properties [127]. However, addition of plasticizers may cause significant changes in the barrier properties of the films, *e.g.* increase film permeability to gases (or aroma compounds), decrease the ability of the film to attract water or decrease the tensile strength [81, 128].

2.3. ANTIMICROBIAL, ANTIOXIDANT AND OTHER FUNCTIONAL AGENTS

The new generation of edible films and coatings is being especially designed to increase theirs functionalities by incorporating natural or chemical antimicrobial agents, antioxidants, enzymes or functional ingredients such as probiotics, minerals and vitamins [5-7]. Antimicrobial and antioxidant coatings have advantages over direct applications of the antimicrobial or antioxidant agents because they can be designed to slow down the diffusion of the active compounds from the surface of the food. Edible coating can enhance the nutritional value of foods by carrying basic nutrients and / or nutraceuticals in its matrix. The sensory quality of coated products can be also improve if in the matrix flavour and pigment are added.

2.3.1. Antimicrobial Agents

Common chemical antimicrobial agents used in food systems, such as benzoic acid, propionic acid, sodium benzoate, sorbic acid, and potassium sorbate, may be incorporated into edible films and coatings to inhibit the outgrowth of both bacterial and fungal cells [129]. In fact, the antimicrobial compounds, when establish contact with food, inhibit the growth of microorganisms present in the surface [130]. However, due to the health concerns of consumers related to chemical preservatives, the demand for natural foods has spurred the search for natural biopreservatives in edible films-forming preparations. Chitosan, which has so far attracted numerous works regarding its antifungal and antibacterial properties is discussed in the above paragraph. The most frequently used biopreservatives for antimicrobial are lysozyme and nisin. These biopreservatives are shown bactericidal effect on Gram (+) bacteria but they can also become effective on Gram (-) bacteria if they are combined with chelating agents such as ethylene diamine tetraacetic acid (EDTA) [131]. Common other biopreservatives that may be used in edible films and coatings are bacteriocins, such as lacticin [132] and pediocin [133] and antimicrobial enzymes, such as chitinase and glucose oxidase [134]. Mitrakas et al. [135] are reported that during the storage at 5 °C, the use of calcium alginate film with antimicrobial as potassium sorbate results in a significant delay in microbial growth on the cylindrical samples of potato, which may help extend the shelf life of an osmo-dehydrated, refrigerated product. Recently, different studies have also been conducted on the use of the enzyme lactoperoxidase (LPS) in edible films and coatings [136]. LPS, that has antimicrobial and antioxidant properties, is found in the mammary, salivary, and lachrymal glands of mammals and in their respective secretions [137]. LPS shows bactericidal effect on Gram-negative bacteria and bacteriostatic effect on Gram-positive bacteria [138]. LPS also presents antifungal and antiviral activities [137]. Natural antimicrobial compounds have also been incorporated into protein or polysaccharide-based matrices: rosemary, and garlic essential oils [139] oregano, lemon grass, or cinnamon oil at different concentrations [66].

2.3.2. Antioxidant Agents

Generally, edible films and coatings have as additive lipids to reduce water vapour transfer due to its hydrophobic character. Therefore, incorporation of antioxidants in edible films-forming preparations to increase product shelf life by protecting foods against oxidative rancidity, degradation, and discoloration is became very popular [11]. The reader can also note that most antimicrobial compounds have antioxidant properties. Natural antioxidants such as phenolic compounds, vitamins E and C in place of synthetic antioxidants are extensively used in edible films. For example, the antioxidants citric and ascorbic acid were incorporated into methylcellulose-based edible coatings in order to control oxygen permeability and reduce Vitamin C losses in apricots during storage [140]. Xanthan gum coatings mixed with vitamin E enhanced nutritional quality and improved the surface colour of peeled baby carrots [59]. Carrageenan or whey protein coatings with added antibrowning agents, ascorbic acid or citric acid, effectively maintained colour of apple slices during storage and extended the shelf-life of minimally processed apple slices by 2 week when stored in packed trays [29]. Recently, banana slices coated with calcium chloride, ascorbic acid and cysteine prevented product weight loss and increase of polyphenol oxidase activity during the 5 days of storage [141]. Essential oils exhibit a wide range of biological effects, including antioxidant and antimicrobial properties. In particular they exhibit antibacterial activity against food borne pathogens. The phenolic components, such as carvacrol, camphor, eugenol, linalool and thymol are most active and appear to act principally as membrane permeabilisers [142]. Oregano or garlic essential oil added whey protein isolate edible films exhibited larger inhibitory zones on *Staphylococcus aureus*, *Salmonella enteritidis*, *Listeria monocytogenes*, *Escherichia coli* O157:H7 and *Lactobacillus plantarum* as compared to rosemary essential oil incorporated into film-forming preparations [139]. Furthermore, antimicrobial films prepared by incorporating oregano oil into sorbitol-plasticized whey protein isolate films was effective in increasing the beef's shelf life by a factor of 2, while minimising changes in colour [143]. Maizura *et al.* [144] reported that starch-alginate edible films containing lemon grass oil are effective in inhibiting the growth of Escherichia coli O157:H7 at all levels. However, generally all these studies showed their efficacy in vitro against various microorganisms but they were not tested with real foods. Moreover, there is a lack of available information about their possible impact on the aroma and flavour of the coated products.

2.3.3. Other Functional Agents

Edible films and coatings a can be used as a host for carring basic nutrients and/or nutraceuticals that are lacking or are present in only low quantity in fruits and vegetables. Recently, a study reported by Tapia *et al.* [23] showed that alginate- and gellan-based edible films seem to be efficient in supporting *Bifidobacterium lactis* Bb-12 on fresh-cut apple and papaya. Indeed, values greater than 10^6 CFU/g of *Bifidobacterium lactis* Bb-12 were maintained for 10 days during refrigerated storage of fresh-cut apple and papaya. However, the bifidus-containing films were more permeable to water vapour than the corresponding cast films. The use of edible films and coatings based on fructooligosaccharides (FOS) showed benefits of FOS like their specific physiological effect on human digestive track especially growth stimulation of beneficial bifidobacteria in the digestive tract, low calorific value and non-cariogenicity [145].

The development of nutritionally fortified edible films and coatings strongly depends on the type of carriers (film-forming materials) and the type and concentration of nutraceuticals added into the film-forming solutions [146]. Nutraceuticals as calcium, zinc, or vitamin E incorporated in a chitosan-based film matrix provided alternative ways to fortify fresh fruits and vegetables [146]. Vitamin E is believed to protect the human body against certain types of cancers, cardiovascular disease, cataracts, arthritis, diabetes, and Alzheimer's disease whereas calcium is a common intracellular messenger, a cofactor for extracellular enzymes and proteins, and also is essential for the development of bone and teeth [147]. The integration of calcium and vitamin E into edible coating around baby carrots improve the nutritional value of this food. This result has also been successfully applied on fresh and frozen strawberries and raspberries [54]. Flavour and pigments agents may also be added to edible coatings to improve the sensory quality of coated foods. However, there is a lack of available information on these applications in the literature.

Chapter 3

3. PROPERTIES OF HYDROCOLLOID EDIBLE FILMS AND COATINGS

3.1. THICKNESS

The thickness of edible films and coatings is an important parameter since it directly affects the biological properties and the shelf life of the coated food. The effectiveness of edible films and coatings for protection of food depends primarily on controlling the spreading of the coating solutions, which affect the thickness of the film.

When a drop of a liquid A is placed on a solid B (for example fruit peel´s or a vegetable) the liquid can wetting totally or partially the solid B. The parameter that distinguishes this two states is the spreading coefficient [148]:

$$S_{A/B} = [E_{solid}]_{dry} - [E_{solid}]_{wet} = \gamma_{solid/air} - \gamma_{solid/liquid} - \gamma_{liquid/air} = \gamma_{liquid/air}(\cos(\theta) - 1) \quad (1)$$

If $S_{A/B} > 0$, the liquid drop spreads completely in order lower its energy, i.e. the contact angle $\theta = 0$ (See Figure 1.B).

If $S_{A/B} < 0$, the liquid drop does not spread but forms at equilibrium a spherical cap resting on the solid with a contact angle θ. For $\theta < 90°$ the liquid is said to be "mostly wetting" and for $\theta > 90°$ "mostly non-wetting" (See

Figure 1.C). Furthermore, by projecting the equilibrium forces onto the solid plane, one obtains Young's relation:

$$\gamma_{liquid/air}\cos(\theta) = \gamma_{solid/air} - \gamma_{solid/liquid}.$$

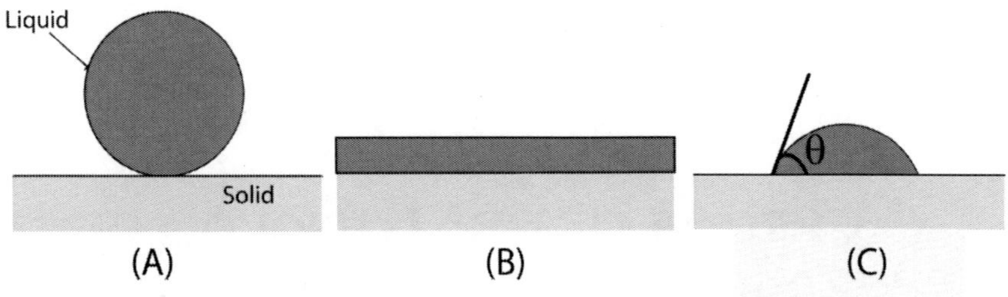

Figure 1. Liquid drop on a smooth solid subtract. (A) No wetting, hydrophobic solid (θ=180º); (B) Total wetting, hydrophilic solid (θ=0); (C) Partial wetting 0< θ<180º.

However, it is difficult to estimate $S_{A/B}$. Indeed, although the interfacial tension between liquid and air $\gamma_{liquid/air}$, can be measured accurately, there are no direct methods for measuring $\gamma_{solid/air}$ and $\gamma_{solid/liquid}$.

A commonly used approach, the Owens-Wendt theory, is to express any surface tension γ (usually against air) as a sum of components due to dispersion forces (γ^d) and polar forces(γ^p) [149]: $\gamma = \gamma^d + \gamma^p$. For example, at 25ºC an air-water interface has a $\gamma^p_{liquid/air} = 50.4$ mJ.m^{-2} and a $\gamma^d_{liquid/air} = 21.5$ mJ.m^{-2}.

The interfacial tension between two phases α and β is then expressed in terms of these two components for each phase:

$$\gamma_{\alpha\beta} = \gamma_\alpha + \gamma_\beta - 2\sqrt{\gamma^d_\alpha \gamma^d_\beta} - 2\sqrt{\gamma^p_\alpha \gamma^p_\beta} \qquad (2)$$

At this point it is convenient to introduce two important works, namely, the ''work of cohesion'' and the ''work of adhesion". The work of cohesion, W_c, is defined as the reversible work required to separate two surfaces of unit area of a single material with surface tension γ whereas W_a, the work of adhesion is defined as the reversible work required to separate unit area of interface between two different materials (See Figure 2). The works are given by:

$$W_c = 2\gamma_{liquid/air} \tag{3}$$

$$W_a = \gamma_{solid/air} + \gamma_{liquid/air} - \gamma_{solid/liquid} \tag{4}$$

The balance between W_c and W_a allows to determine the spreading coefficient $S_{A/B}$, according to:

$$S_{A/B} = W_a - W_c \tag{5}$$

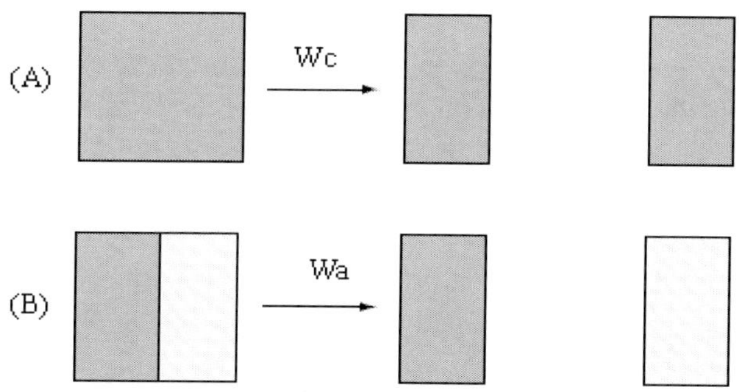

Figure 2. (A) Work of cohesion; (B) Work of adhesion.

Thus, using Eq. 2 the work of adhesion can be express:

$$W_a = W_a^d + W_a^p = 2\left(\sqrt{\gamma_{solid/air}^d \gamma_{liquid/air}^d} + \sqrt{\gamma_{solid/air}^p \gamma_{liquid/air}^p}\right) = \gamma_{liquid/air}\left(\cos(\theta) + 1\right) \tag{6}$$

Rearranging Eq. 6, yields:

$$\frac{1+\cos(\theta)}{2} \frac{\gamma_{liquid/air}}{\sqrt{\gamma_{liquid/air}^d}} = \sqrt{\gamma_{solid/air}^p} \sqrt{\frac{\gamma_{liquid/air}^p}{\gamma_{liquid/air}^d}} + \sqrt{\gamma_{solid/air}^d} \tag{7}$$

It is possible to obtain $\gamma^{p}_{solid/air}$ and $\gamma^{d}_{solid/air}$ by plotting the experimental data in a graph of $\frac{1+\cos(\theta)}{2}\frac{\gamma_{liquid/air}}{\sqrt{\gamma^{d}_{liquid/air}}}$ vs $\sqrt{\frac{\gamma^{p}_{liquid/air}}{\gamma^{d}_{liquid/air}}}$. The reader can find various data $\gamma^{p}_{liquid/air}$ and $\gamma^{d}_{liquid/air}$ in [150]. Since $\gamma_{solid/air}$ is the sum of $\gamma^{p}_{solid/air}$ and $\gamma^{d}_{solid/air}$, the parameter $\gamma_{solid/liquid}$ can be determined by the Eq. 2. Thus, the important spreading coefficient $S_{A/B}$ can be calculated.

For a solution of pure starch or carrageenan coating a strawberry fruit *S*~-45mJ.m^{-2} whereas for a chitosan solution *S*~-38mJ.m^{-2} [33]. In all cases, the thickness of the edible film will be thick since the liquids do not spread.

However, films and coatings must not exceed a critical thickness to no reduce drastically the internal O_2 concentration and to increase CO_2 concentration from anaerobic fermentation. Coating thickness depends on the physical properties of the solution (such as density, viscosity, and surface tension) but also of the method used in the film formation: dipping, spraying solvent casting, skimming, etc. [109]. The methods used for the film formation are discussed in the paragraph 4.

The method to measure the film or coating thickness can be classified in two groups: the contact method and the non-contact method. The non-contact method is more effective than the contact method since this technique is non-destructive for the sample. Among the contact methods, due to its simplicity, the direct metric measurement is most frequently used. The coating is peeled from the food surface and measured using a micrometer. Usually, a film has thickness inferior to 0.25mm or 0.001 inch. A thickness greater than 0.25mm is called a sheet.

Common non-contact methods to analyze coating thickness directly from the surface of coated foods include optical or scanning electron microscope, confocal Raman microspectrometry (CRM), Fourier transform (FT)-Raman spectrometer and surface enhanced Raman scattering (SERS) [151-154]. A representative example of the non-contact method is the optical method, which is based mainly on light interference [155]. The interference method can usually measure only a relative thickness by means of interference fringe. However, a method for an absolute measurement has recently been developed by means of multi-wave laser interference at many incident angle [153]. Two methods are proposed: one can be used when an index of refraction of the film, a wavelength λ; and two successive angles of incidence at which the sinusoidal light intensity has minimum values, are known (Method I), and another can be used without an index of film refraction when three successive angles of incidence and a wavelength are known (Method II). The smallest measurable thickness is 1.43 λ for Method I, and 2.5 λ for Method II. The largest measurable thickness is about 100 λ for both methods [153]. The simultaneously measure of the thickness and of the surface profile of single film as well as multiple-layer film can be obtained using the interferometry in combination with confocal microscopy [156-157]. However, when a food is coated by dipping, different theoretical approaches can be used to estimate film thickness from coating solution properties [158-159]. Cisneros-Zevallos and Krochta [109] used hydroxypropyl methylcellulose to coat Fuji apples by dipping. They reported that coating thickness varied with viscosity, concentration, density, and draining time of the biopolymer solution.

3.2. MECHANICAL PROPERTIES

Typically, the mechanical resistance of hydrocolloidal films was studied according to three parameters: tensile strength (*TS*), Young's modulus (*Y*) and percent of elongation at break (*E*). Tensile strength, elongation properties, Young's modulus are determined from stress–strain curves according to the norm of the American Society for Testing and Materials (ASTM).

In Figure 3, a strain-stress curve is presented. TS is calculated by dividing the maximum load on the film before failure (σ_Y) by the initial cross sectional area (*S*). *E* is expressed as the percentage of change in the original length of film before breaking and gives a measure of the extensibility of the films (ε_B). *Y* is experimentally determined from the slope of a stress-strain curve created during the tensile test ($Y = \frac{\sigma_L}{\varepsilon_L}$) which is a measure of film stiffness [160].

In the literature some results of mechanical properties (*TS*, *E*, *Y*) of biopolymer and synthetic films are reported [160-161]. However, due to potential difference in manipulation or testing procedures, it was very difficult to compare the results. Values of *TS* and *E* are reported in the Table 2 for various polymer films. For polysaccharides films, *T*S values vary between 10 and 100 MPa whereas the *E* values range between 1 and 80%. Generally, protein films have a *TS* lower while theirs *E* are higher [127]. Comparison with common synthetic polymers used in films-forming preparations, *TS* values approach those of polysaccharides films: 9 to 20 Mpa for low-density polyethylene (LDPE) and 10 to 60 MPa for high-density polyethylene (HDPE). However, LDPE and HDPE have higher *E* values (up to 1000%), which are greater than those of protein films. Thus, it is necessary to add plasticizers into polysaccharides or proteins film-forming preparations to increase *E* values. However, increasing concentration of these plasticizers resulted in decreased tensile strength (*TS*). For example, films produced with low molecular weight chitosan at 3% w/w in 1% acetic acid, with glycerol as a plasticizer at 0.25 and 0.50 mL/g of chitosan, were reported to have tensile strength (TS) of 15 to 35 MPa and percent elongation at break (%E) of 17 to 76 [162].

Lipids incorporated into film-forming preparation increase the hydrophobicity of edible films but also generally cause a weakening of film strength [37-163].

Table 2. Tensile strength (*TS*) and elongation at break (*E*) for various polymer films. All data are issue of [164] excepted a) [165], b) [166] and c) [2]

Polymer		Tensile Strength (TS) Mpa	Elongation at break (E) %
Cellulose	acetate	44 - 64	20 - 50
	Hydroxypropyl cellulose [c)]	14 - 15	30 - 205
	Methylcellulose [c)]	55 - 60	10 - 20
Collagen		1 - 70	10 - 70
Chitsan [a)]		10 - 100	20 - 80
Gelatin	Type A	25 - 85	7 - 22
	Type B	28 - 140	7 - 12
Starch	Waxy maize	34.9	1.7
	Tapioca	44	3.4
	Potato	44.2	3.1
	Wheat	46.3	2.9
	Corn (A)	46.4	2.5
	Corn (B)	46.7	3.2
Polyethylene	linear low density	9 - 20	100 - 1200
	linear high density	10 - 60	400 - 1800
zein [b)]		3 - 4	50 - 120

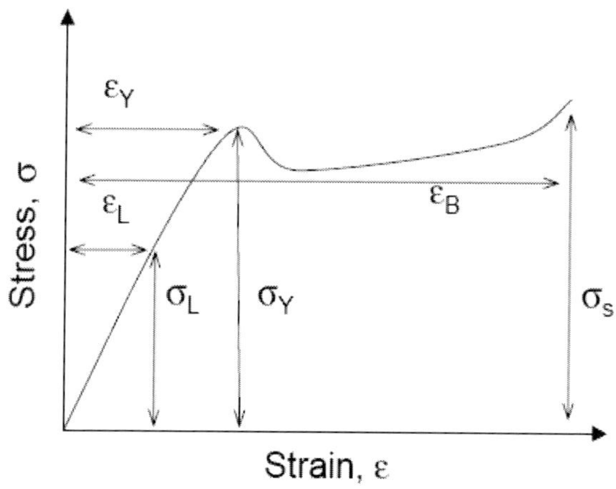

Figure 3. Stress–Strain behaviour of a hydrocolloid film. The explanation of all symbols are given in the text.

3.3. Barrier Properties

When an edible film is applied on the food, the surface gas concentration may change during storage. The chemical composition of the food surface is dynamic and these changes occur mainly due to food metabolism, microbial respiration, gas solubility and permeability of the edible film. Food and microbial metabolism are responsible for consumption of oxygen and CO_2 production. The microbial activity can affect the composition of oxygen and CO_2, however this is only significant when it has reached the end of shelf life and deterioration is evident [167]. Therefore, the efficiency of an edible coating strongly depends on its barrier properties to gas, water vapor, aroma and oil which in turn depends on the chemical composition and structure of the coating-forming polymers, the characteristics of the product, and the storage conditions.

3.3.1. Gas Barrier

Coatings are used to create a controlled or modified atmosphere inside the fruits and vegetables that will delay ripening and senescence in a manner similar to the most costly controlled or modified atmosphere storage. The transport of gas molecules in an edible film consists of 3 steps (See Figure 4): adsorption of the permeant onto the edible film surface, diffusion of the permeant from one side of the edible film to the other side, and then the desorption of the permeant from the edible film.

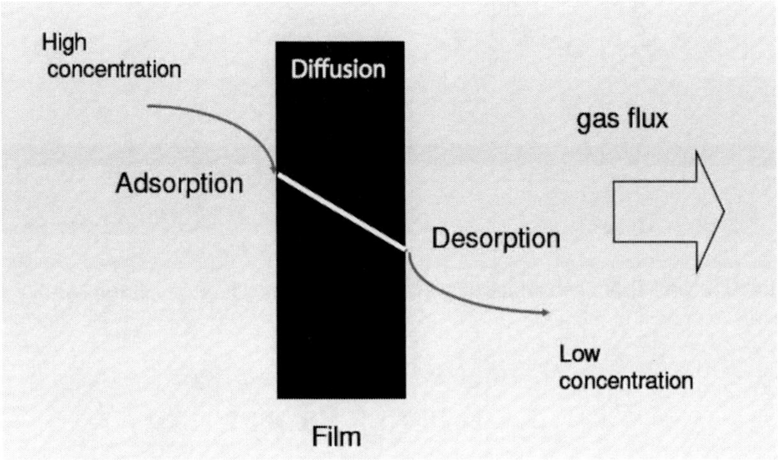

Figure 4. Transport of gas molecules in an edible film.

Diffusion of simple gases (e.g., 0_2, N_2, CO_2) through a film generally obeys Fick's law, i.e. [168]:

$$J = -D\frac{\partial C}{\partial x} = D\frac{C_1 - C_2}{l} \tag{1}$$

Where: l is the thickness of the edible film, D the diffusion coefficient and C the concentration.

Moreover, the Henry's law says that the concentration of the gas can be expressed as:

$$C = \Sigma p \tag{2}$$

where Σ is the solubility coefficient of the gas in the membrane, and p is the pressure of the gas. The solubility, diffusivity, and kinetic diameter data of N_2,

O_2 and CO_2 gas in water at 25°C are presented in the Table 3. The kinetic diameter of CO_2, O_2 and N_2 are very close. However CO_2 is much more permeable (product of diffusivity by the solubility coefficient) than the two other gases because of its much higher solubility in moist films due to presence of water in the films. Furthermore, the solubility and diffusivity of O_2 are larger than N_2. Therefore, in the case of a polar film matrix, these characteristics attribute to the permeability in the order of $CO_2 > O_2 > N_2$.

Table 3. Solubility, diffusivity, and kinetic diameter data of N_2, O_2 and CO_2 gas in water at 25°C [169]

Gas	Solubility Coefficient (mol.(L.atm)$^{-1}$)	Diffusivity 10^{-9}(m^2/s)	Kinetic diameter (10^{-10} m)
N_2	$6.1x10^{-4}$	2	3.65
O_2	$1.3x10^{-3}$	2.42	3.46
CO_2	$3.4x10^{-2}$	1.91	3.3

Furthermore, what happens inside the edible film is not relevant to the design, therefore the concept of permeability can be obtained by combining the Fick's first law (Eq. 1) with Henry's law (Eq.2):

$$J = D\Sigma \frac{p_1 - p_2}{l} \quad (3)$$

The product of diffusion coefficient and solubility coefficient is equal to the permeability coefficient, $\Pi = D\Sigma$, which characterizes the intrinsic permeability of the edible film.

For a *i*-multilayer edible-film, the flux *J* will be described by:

$$J = \frac{p_1 - p_2}{\sum_i \frac{l_i}{\Pi_i}} \quad (4)$$

It is important to point out that in respiring-foods the permeability of the edible-film is extremely relevant since foods consume O_2 to produce an equilibrium condition. Oxygen uptake by a food often results in deleterious reactions which affect its flavor, nutritional quality and acceptability. The oxygen uptake for the food (R) is usually estimated by a Michaellis Menten model [170]:

$$R = R_{max}\left(\frac{P_{O_2}}{K + P_{O_2}}\right) \tag{5}$$

Thus, it is possible predict easily the equilibrium partial pressure of oxygen when $R = J$ with the following equation:

$$V_F R_{max}\left(\frac{P_{O_2}}{K + P_{O_2}}\right) = \frac{A_F}{\sum_i \frac{l_i}{\Pi_{O_2,i}}}\left(p_{O_{2,AIR}} - p_{O_2}\right) \tag{6}$$

Where

V_F and A_F are volume of food and area of food, respectively.

In addition, this equation can be written with dimensionless numbers (α and β):

$$\alpha\left(\frac{\beta}{1+\beta}\right) = \left(\beta_{AIR} - \beta\right) \tag{7}$$

where

$$\alpha = \frac{V_F}{A_F} R_{max} \sum_i \frac{l_i}{\Pi_{O_2,i}}, \ \beta = \frac{P_{O_2}}{K} \text{ and } \beta_{AIR} = \frac{P_{O_{2,AIR}}}{K}$$

Therefore the partial pressure of oxygen in contact with the food will be minor that air (<0.21 atm):

$$\beta = \frac{1}{2}\left(\left(\beta_{AIR} - \alpha - 1\right) + \sqrt{\left(\beta_{AIR} - \alpha - 1\right)^2 + 4\beta_{AIR}}\right) \tag{8}$$

This dimensionless can be used to estimate the reduction in the oxygen uptake in comparison with a food respiring without edible-film (in contact with air):

$$\left(\frac{R}{R_{AIR}}\right)=\frac{R_{MAX}\left(\frac{p_{O2}}{K+p_{O2}}\right)}{R_{MAX}\left(\frac{p_{O2,AIR}}{K+p_{O2,AIR}}\right)}=\frac{\beta(1+\beta_{AIR})}{\beta_{AIR}(1+\beta)} \tag{9}$$

Figure 5 shows the relationship between the reduction of respiration (R/RAIR) and the edible-film to different and β values.

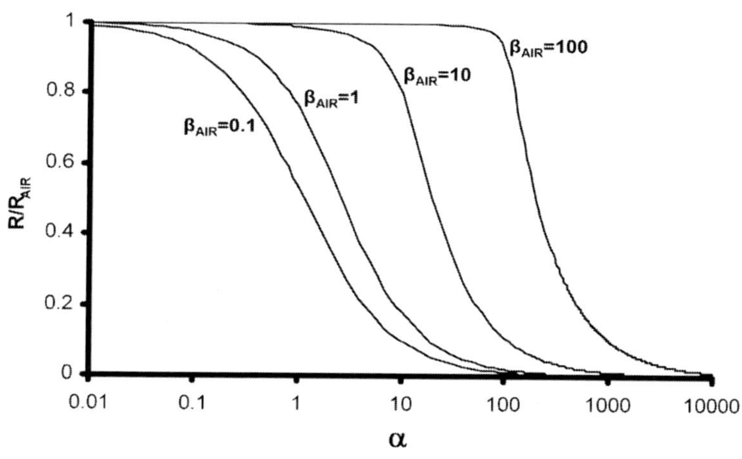

Figure 5. Reduction of respiration (R/R_{AIR}) and the edible-film to different α and β values.

It is important to point out that in respiring-foods foods consume O_2 to produce CO_2, which must be released to the environment to produce an equilibrium condition [171]. If the CO_2 concentration is higher, this gas could produce an inhibition on respiration. The uncompetitive enzymatic model is usually used to predict this effect:

$$R=R_{MAX}\left(\frac{p_{O2}}{K+p_{O2}\left(1+p_{CO2}/K_{I}\right)}\right) \tag{10}$$

The CO_2 production is obtained with the respiration quotient (Q). Typically, Q varies between 0.8 and 1.5 [170]. Then, to predict the new equilibrium conditions is necessary to solve a 2 x 2 system equations:

$$V_F R_{\text{MAX}}\left(\frac{p_{O2}}{K + p_{O2}(1 + p_{CO2}/K_I)}\right) = \frac{A_F}{\sum \frac{L_i}{\wp_{O2,i}}}(p_{\text{O2,AIR}} - p_{\text{O2}})$$

$$V_F \text{Q} R_{\text{MAX}}\left(\frac{p_{O2}}{K + p_{O2}(1 + p_{CO2}/K_I)}\right) = \frac{A_F}{\sum \frac{L_i}{\wp_{CO2,i}}}(p_{\text{CO2}} - p_{\text{CO2,AIR}})$$

3.3.2. Water Vapour Permeability (WVP)

In the case of water vapour diffusing through a polar film matrix, J will increase as Δp is moved up in the vapour pressure spectrum. This occurs because the adsorption isotherms for hydrophilic polymers are sigmoïdal, causing Σ to increase significantly when the vapour pressure rises. The greater amount of adsorbed water plasticizes more the film causing the diffusion coefficient D to increase. Hence, Fick's law is not strictly obeyed and D is not a constant for water vapour migration through hydrophilic film matrices [10].

The WVP is calculated by the following equation [172]:

$$WVP = \frac{\Delta m}{\Delta t}\frac{l}{A\Delta\Delta} \tag{1}$$

The WVP correspond to the amount of water vapour (Δm) transferred through a film of area (A), of thickness l during a finite time (Δt). The ΔP is the vapour pressure difference across the film.

Published WVP values for edible films should be cautiously compared because of the difference in plasticizers, temperature, relative humidity gradient, etc. The increase of the film permeability with thickness indicates a water affinity of the film that could be attributed to hydrophilic compounds. Then it seems that the influence of thickness varies with the lipid nature or the film composition [37].

3.3.3. Aroma Barrier

An understanding of sorption between flavour compounds and edible film matrix requires knowledge of the chemical and physical structures of both the flavour compound and the polymer. [173]. Studies on the aroma permeability of hydrocoloid edible films are very scarce. The flavour diffusion coefficient D in edible films are in the order of 10^{-13} m^2 s^{-1} [1, 173]. Flavour absorption may affect the flavour of a food as well as the mechanical properties of the edible film, such as tensile and heat seal strength and permeability, or cause delamination of the polymeric structure [174]. Finally, an example of study aroma permeability was reported by Miller and Krochta [175]: a whey protein film was better than Vinylidene chloride copolymer (co.VDC) and was comparable to ethylene vinyl alcohol copolymer (EVOH) as barrier to limonene (citrus arome).

3.3.4. Oil Barrier

Film coatings may become a good alternative to reduce oil uptake during frying. Several hydrocolloids with thermal gelling or thickening properties, such as proteins or carbohydrates, were tested to reduce oil migration [176-178]. Hydrocolloid films are excepted to be oil-resistant due to their substantial hydrophilicity. García *et al.* [178] reported that, in deep-fat frying potato strips and dough discs, methylcellulose coatings are more effective in reducing oil uptake than hydroxypropylmethyl cellulose. Sorbitol addition was necessary to maintain coating integrity and improve barrier properties. Some potentiality and applications of edible coatings as an oil barrier are presented by García *et al.* [179].

3.4. Optical Properties: Gloss, Transparency and Colour

In the last years, research efforts have been made to improve mechanical and barrier properties of edible films and coatings but few studies have focussed on optical properties such as colour, gloss and transparency. Optical properties are characteristics of surfaces which are detected by human vision affecting some crucial aspects of food quality. The internal and surface film microstructure plays an important role in optical properties of the film. Indeed, the intensity of light reflected by the coated food is determined both by the light directly reflected at the interface between air and the coated food surface (specular reflection), and by both the light re-emitted out of the surface in all directions after penetrating into the coating of the food and scattering internally (indirect reflection). Specular reflection is related to the gloss of the coating. According to the ASTM standard D-523, it is measured at 20, 60 and 85° angles from the normal to the coating surface using a flat surface gloss meter. Polished black glass with a refractive index of 1.567 possesses a specular gloss of 100%. The gloss of edible coatings is affected by coating microstructure in particular, by the type and level of surfactant, the distribution and the size particle diameter of dispersed phase, the relative humidity, the storage time and the surface roughness [180-183].

The transparency of films was measured through the surface reflectance spectra in a spectrocolorimeter. A good method to determine the transparency is the Kubelka–Munk theory [184]. This theory models the reflected and transmitted spectrum of a coloured layer based on a material dependent scattering and absorption function. The assumptions of this theoretical model are:

(1) A translucent colourant layer on the top of an opaque background,
(2) Within the colourant layer both absorption and scattering occur,
(3) Light within the colourant layer is completely diffuse,
(4) Geometry should be diffuse illumination and diffuse collection.

The light flux passes through the layer and it is affected by the absorption coefficient K and the scattering coefficient S. The ratio known, as the Kubelka-Munk function, shows that the reflectance of an opaque layer depends only on the ratio of K to S and not on their individual values:

$$\frac{K}{S} = \frac{(1 - R_{\infty})^2}{2R_{\infty}}$$

where

R_∞ is the reflectance of an infinitely thick layer of the material: $R_\infty = a - b$.

With $a = \frac{1}{2}\left(R + \frac{R_0 - R + R_g}{R_0 R_g}\right)$ and $b = \sqrt{a^2 - 1}$. R_0 is the reflectance of the layer with an ideal black background; R the reflectance of the sample layer backed by a known reflectance R_g.

The internal transmittance in the films is quantified by applying: $T_i = \sqrt{(a - R_0)^2 - b^2}$.

The color of edible films are evaluated by a colorimeter or a spectrophotometer. The parameters L (Luminosity), a (Red–Green) and b (Yellow–Blue) are the edible film colour values in the Cielab scale. A white disk (L_0, a_0 and b_0) was used like standard colour. From these values, Chrome ($C = \sqrt{a^2 + b^2}$), Hue ($H = \arctan\left(\frac{b}{a}\right)$), Whiteness Index ($WI = 100 - \sqrt{(100 - L)^2 + a^2 + b^2}$) and difference of colour $\Delta E = \sqrt{(\Delta L)^2 + (\Delta a)^2 + (\Delta b)^2}$ can be calculated. Where $\Delta L = L - L_0$, $\Delta a = a - a_0$ and $\Delta b = b - b_0$.

3.5. Thermodynamic Parameters Associated to Sorption Phenomenon in Edible Films

The interactions of between the sorbant and solvent can define the efficiency of the edible film to control the gas permeability and therefore the stabilities of the food. The relationship between the total moisture content and the water activity (A_w), over a range of values, at constant temperature, yields the sorption isotherm when expressed graphically [185]. When obtained at different temperatures, the energy associated to moisture interactions at low water vapor pressures can be estimated. Various mathematical models have been developed to describe sorption isotherms. GAB equation [186-187] (Eq. 1) has proven to be one of the more versatile to represent sorption values over a wide range of A_w [188-191]. Furthermore its parameter can give information related to the sorption energies involved [198].

$$M = \frac{m_o CK \frac{p}{p_o}}{\left(1 - K\frac{p}{p_o}\right)\left(1 - K\frac{p}{p_o} + CK\frac{p}{p_o}\right)} \tag{1}$$

where

p/p_o = partial water vapor pressure in the food system, m_o = monolayer value (all sorption sites occupied by one molecule), C = constant (related to the net heat of sorption at the monolayer) K = constant (related to the heat of sorption at the multilayer).

The effect of temperature on sorption isotherms can be represented by the Clausius-Clayperon relation [192] (Eq. 2). Graphically, this equation produces the parameter q_{st}, net isosteric heat of sorption or the difference between the total heat of sorption (Q_{st}) and the heat of vaporization of water [193]. This parameter accounts for the excess in energy for removal of water [192].

$$q_{st} = -R\frac{\partial \ln(p/p_o)}{\partial(1/T)} \tag{2}$$

where

q_{st} = net isosteric heat of sorption ($Jg^{-1}mol^{-1}$), p/p_o = partial water vapor pressure in the food system, T= temperature (K) and R = universal gas constant (8,314 $J\ K^{-1}\ mol^{-1}$).

Further analysis of the sorption data can help to elucidate information related to the thermodynamic involved in sorption phenomenon. It can allow the interpretation of the experimental results in accordance with the statement of the theory [194] giving insight of the energies involved in the interactions between sorbant and sorbates. These functions include, with the heat of sorption, the variations of enthalpy and entropy. It has been suggested that enthapy change (*ΔH*) provides indications of the energy variation occurring on mixing water molecules with sorbents during sorption process [195]. Variations in entropy (*ΔH*) might be associated with spatial arrangements occurring in the water-matrix interface in a well defined state [196].

The enthalpy-entropy compensation theory can be applied to evaluate the physical and chemical processes such as sorption phenomenon [194]. This theory states that to reduce free energy during sorption a linear function occurs relating the variations of enthalpy to variations in entropy in the system [194]. The enthalpy-entropy compensation theory has been used to investigate the sorption phenomenon of various biological materials such as cottonseed [194], tropical woods [196] and starch based materials [195].

When complex polymeric structures are related chemically, changes in the enthalpy and entropy can occur. If there is an increase in the intermolecular forces, it can generate a reduction in the configurational freedom of the system [197]. The relation between the difference of enthalpy and the differences in entropy during the sorption process can be mathematically represented by the following equation [195]:

$$\ln Aw = -\Delta H/(RT) + (\Delta\Delta/R) \tag{3}$$

where

A_w = water activity in the food system, T= temperature (K) and R = universal gas constant (8,314 J mol^{-1} K^{-1}). The enthalpy-entropy compensation theory should follow a linear relationship as indicated by Eq. 4 that considers the variation in enthalpy, entropy and Gibbs free energy [193].

$$\Delta S = \frac{(\Delta\Delta + \Delta G)}{T_\beta} \tag{4}$$

where

β= isokinetic temperature (K), ΔH is the change of enthalpy, ΔS is change in entropy and ΔG is the variation in the Gibbs free energy. In order to validate the existence of a true compensation during the sorption process [199-200] proposed a comparison between T_β and the harmonic temperature (T_{hm}), defined in the next function:

$$T_{hm} = \frac{n}{\sum_{i=i}^{n}(1/T)} \tag{5}$$

Where: n= number of isotherms and T = temperature of equilibrium (K). The parameter T_β can be estimated in terms of an interval at (1-α)100% confidence using the following equation:

$$\beta_{m-2,\alpha/2}\sqrt{Var(T_\beta)} \tag{6}$$

Where

$$-\overline{\Delta S})^2 \tag{7}$$

and

$$-\overline{\Delta S})^2 \tag{8}$$

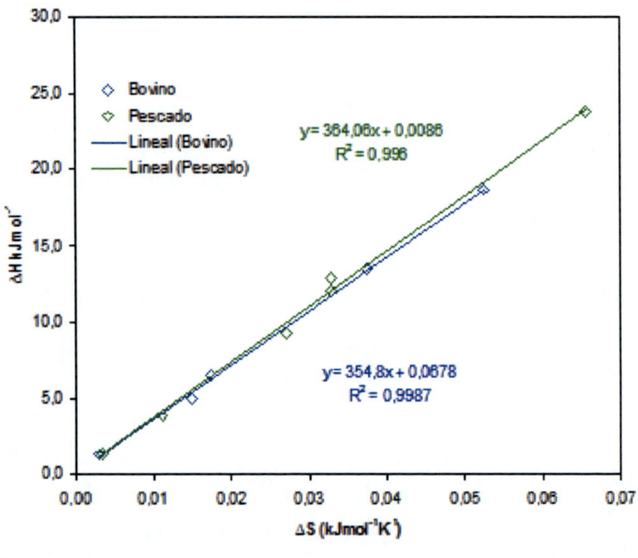

Figure 6. ΔH versus ΔS calculated for fish and bovine gelatin films obtained at 10, 20 y 30 °C. The equations represent the estimated linear regressions for each data set.

In Eq. 6, *m* is the number of ((ΔH), (ΔS)) data pairs, *t* is the *t* value at (m-2) degree of freedom, $\overline{\Delta H}$ is the average differential enthalpy (kJ mol^{-1}) and $\overline{\Delta S}$ is the average differential entropy (J mol^{-1} K^{-1}) [195]. According to [199-200] the compensation theory only exists if $T_\beta \neq T_{hm}$.

Figure 6 depicts the calculated values of ΔH and ΔS by equation 3 using the sorption data obtained from bovine and fish gelatin equilibrated under different relative humidities at 10, 20 and 30°C. This graph clearly shows a linear relationship between ΔH and ΔS ($R^2 > 0.99$) indicating the existence of compensation between both thermodynamic parameters.

The isokinetic temperature (Tβ) and ΔG were calculated using Eq. 4. The expression of Tβ in terms of an interval (95% confidences) was obtained using Eq. 6, 7 y 8 (Table 4).

Table 4. Parameters ΔG, T_{hm} y T_β obtained for from fish and bovine gelatin films

Source	ΔG ($kJmol^{-1}$)	T_{hm} (K)	T_β (K)
Fish	0,0086	292.9	364,1 ± 62.1
Bovine	0.0678	289.5	354,8 ± 19.8

(*) t value for n= 6 and α= 0.025).

The harmonic mean temperature (T_{hm}) was estimated using equation 5, giving values of T_{hm} ~ 292.9K and T_{hm} ~ 289.5K for fish and bovine gelatin films validating respectively. These values were significantly different to the obtained T_β as indicated in table 1, supporting the application of the compensation concept of moisture sorption by theses materials [199-200].

Leffler [201] proposed that if $T_\beta > T_{hm}$, the sorption process is enthalpy controlled and if $T_\beta < T_{hm}$ sorption would be entropy driven. In this work, T_β for both materials were significantly higher than T_{hm} (Table 4), indicating that the sorption mechanism can be considered enthalpy driven.

Chapter 4

4. Usual Methods for Hydrocolloid Films and Coatings Formation

The conditions of the film and coating process have a significant effect on the physical properties of the resulting film. Principal technologies for producing edible films are similar to those for thermoplastic structures: solvent casting and extrusion. Obviously, the conditions are different, but the principles are allied. For the coating the main techniques are: spraying and dipping. Edible films and coating must be uniform and free from defects for optimize their functionalities. In Figure 7, three film defects are presented. The non uniformity, the insertion of bubble and the mechanical damage of the food.

4.1. Solvent Casting

Solvent casting is the most used techniques to form hydrocolloid edible films. Water or water–ethanol solutions or dispersions of the edible materials are spread on a suitable substrate (See paragraph 3.1) and later drying. During drying of the solution, solvent evaporation leads to a decrease in solubility of the polymer until polymer chains align themselves to form films. The choice of the substrate is important to obtain films, which can be easily peeled without any damage after that the solvent is evaporated. Generally, the film are air-dried during several hours in the ventilated oven [202]. Infrared drying chambers are advantageous in that they hasten the drying process [203]. Optimum moisture content (5–10% w/v) is desirable in the dried film. Film structures depend on the drying conditions (Temperature and relative humidity), wet casting thickness as well as the composition of the casting solution. Films prepared from pure polymers tend to be brittle and often crack upon drying therefore an addition of plasticizers is necessary [204]. In general, rapid drying of cast polymer solutions limits the development of intermolecular associations within the film structure since the mobility of the polymer chains are restricted when the solvent concentration decrease [160]. Thus, a rapid drying often causes defects such as cracks or curling in the films [205]. Alcantara *et al.* [206] reported that film formation from WPI with a rapid drying at 95ºC were thinner, stronger and had a lower E than films dried more slowly at 21ºC. Polysaccharide–lipid multilayer films with water vapour barrier have been also produced by casting [38-207]. The film forming solutions with a mixture of gelatine, while still hot, are cast by pipetting solutions onto acrylic or flat glass plates and dried at room temperature [208]. Microwave drying was also been investigated and compared with air-drying by Kaya and Kaya [209]. They reported that the film dried by microwave were less hazy and glossier than film dried at room condition.

4.2. Extrusion

In the plastics industry, the most synthetic polymer films are produced by extrusion. The application of extrusion technology to edible films, such as starch films, is reported by Finkenstadt and Willett [210] or Guan and Hanna [211]. The extrusion process is based on the thermoplastic properties of polymers when plasticized and heated above their glass transition temperature under low water content-conditions. The extrusion processing of polymer products often need the addition of plasticizers such as polyethylene glycol, sorbitol in an amount of 10% to 60% w/w (see paragraph 2.2). In comparison with the solvent casting method, the extrusion method is attractive for industrial processes since it does not require solvent addition and a time of evaporation. Pectin has also been combined with glycerol and extruded into films [115, 212]. Furthermore, the formation of extruded protein films produced several scientific patents. For example, Naga *et al.* [213] developed extruded films of soy protein with a plasticizers (polyhydric alcohols) and water whereas Koepff *et al.* [214] reported extruded gelatins films, smooth, uniform, transparent with slight opacity. Such films become increasingly flexible with increased hydration.

To form multilayer films the technique of co-extrusion may be used. However, differences in the physical and chemical properties of the constituent materials in such multilayer films can lead to formation of a number of defects. These defects may be detrimental in terms of the eventual mechanical, barrier, and optical performance of the films. Recently, a defect, the zigzag (or viscous fingers) at the interface of extruded films was evaluated based on steady-state calculations [215].

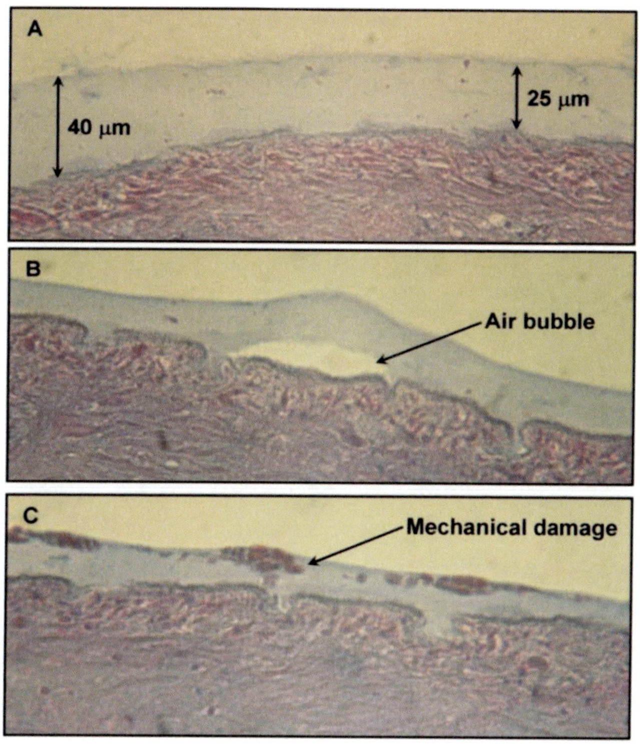

Figure 7. Histochemistry technique showing the cross-section of the alginate coating on clams and the defects: (a) non-uniformity, (b) insertion of air bubble and (c) mechanical damage of the food product. The coating was made by dipping.

4.3. DIPPING

Nowadays, dipping is a common method for applying coatings on fruits and vegetables [7]. The coating by dipping on the coating solution properties such as density, viscosity and surface tension, as well as food withdrawal speed from the coating solution [109]. Different theoretical approaches can be used to estimate film thickness from coating solution properties [158-159]. Generally the food was dipped into the film-forming solutions between 5 and 30 s [55,140].

Direct application of antimicrobial agents onto food surfaces by dipping has proven to be less effective as there is loss of activity because of leaching onto the food, enzymatic activity, and reaction with other food components [216]. However, the brushing method for the application of film solution to fresh beans and strawberries was found to be better than wrapping and dipping methods in terms of reducing the moisture loss [140].

The application of edible coatings to minimally processed fruits is not easy because it is difficult to obtain a good adhesion of the coating to the hydrophilic surface of the cut fruit. A solution to solve this problem could be a multilayer technique: the layer-by-layer electrodeposition where two or more layers of material with nanometer dimensions are physically or chemically bonded to each other [7, 217]. In Figure 7, the layer-by-layer technique is presented to coat highly hydrophilic food systems such as fresh-cut of apple fruit.

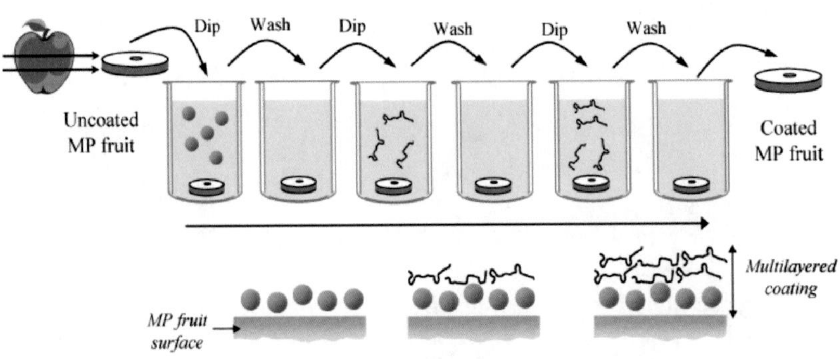

Figure 8. Schematic representation of coating a minimally processed (MP) fruit with a multilayered edible coating by using the layer-by-layer assembly with three dipping and washing steps (Reproduced with permission from [7]).

4.4. SPRAYING

In the food industry, coating by spraying is the conventional method generally used when the coating-forming solution is not very viscous. Indeed, highly viscous solution cannot be or very uneasily sprayed, and thus only dipping techniques can be apply, giving high thickness to the coating.

Initially, in the case of plan jet, the spray exists the nozzle as a two dimensional sheet of fluid. This sheet then breaks up into a series of cylindrical ribbons which eventually undergo the same type of breakup as laminar capillary jets [218]. The control of the final drop size and therefore of the quality of the coating depend on spray gun and nozzle, temperature, air and liquid flow rate, humidity of incoming air and of the polymer solution [219-220]. Nowadays, programmable spray systems are available for automation during such operations. Classic spraying system can produce a fine spray with relatively drop-size distribution up to 20 μm whereas electrospraying has shown to be capable of producing uniform particles of less than 100 nm from polymer and biopolymer solutions. Futhermore, other different factors are critical in the formation of polymeric films by spraying systems, such as drying time, drying temperature, drying method, etc.

There are some latent diseases in fruits, which usually occur in postharvest storage. Therefore, studies were focused on the effects of preharvest chitosan spray or/and postharvest chitosan coating to extended the protective effect against decay of fruit and to avoid the use of chemical products [221]. Recently, Meng at al. [222] are reported that the treatment with low chitosan concentration (1 g L^{-1}) before harvest and a coating of the fruit with a higher concentration (10 g L^{-1}) after harvest have a beneficial effect on fruit quality and resistance to fruit decay. A similar effect was reported by Bhaskara Reddy *et al.* [223] on strawberry fruit.

Chapter 5

5. Future Needs in Edible Film and Coatings

5.1. Scale-Up of Edible Film and Coating in a Industrial Production

Despite the numerous published studies corroborating that hydrocollloid edible films and coatings are economical (as starch...), extend shelf life and enhance quality and microbial safety of food, commercial applications are still very limited. Generally the works report experimental set-up built at the laboratory scale and a coating process made by hand. Moreover some techniques are unsuitable for the industry. For example, the production of film by solvent casting is quite costly, time consuming (up to 48h) and the drying may generate, for the food product, surface dehydration and moisture loss. Spraying, due to the large droplet velocity, may cause damage to a delicate food product such as berries. Further research on the application of hydrocolloid coatings at the real scale is needed, in particular for resolving the problems of non-uniformity in the thickness and in the homogeneity of film.

5.2. Control of Microstructures for Edible Films

To keep the food product attractive for the consumers, the surface of edible films and coatings must appear uniform and without defects. However, the classic methods such as dipping, spraying, solvent casting not allow to control the film thickness. Moreover, most edible films and coatings are formed with little consideration of the internal structure created whereas when functional ingredients (antimicrobials, antioxidants, colour agents, emulsifiers, flavour, nutraceuticals) are incorporated, the film lost its homogeneity.

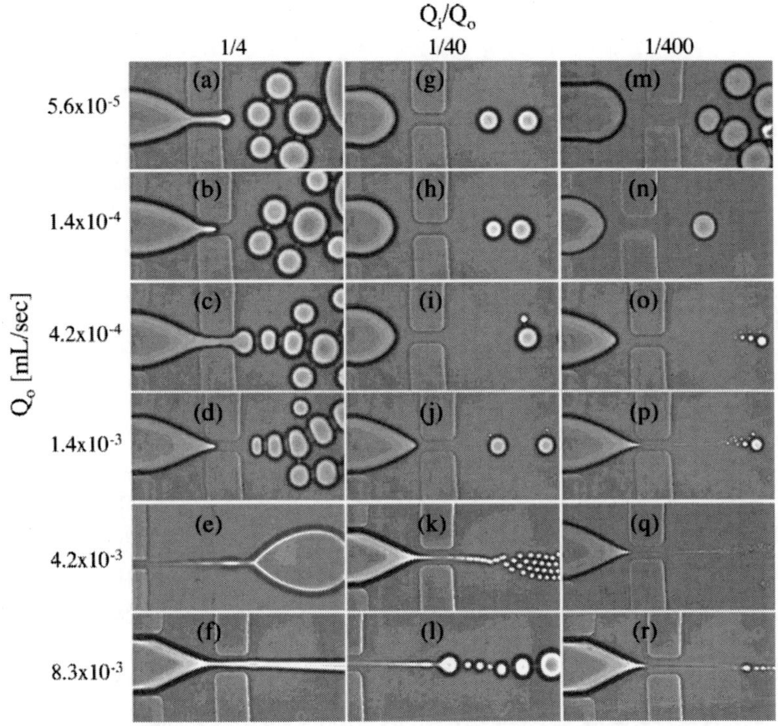

Figure 9. Micro device having a "flow-focusing orifice" before the outlet microchannel. Each image shows the drop size and pattern of drop stream at specific values of Q_o (rows) and Q_i / Q_o (columns), Q_i and Q_o being the flow rates of water and oil, respectively (Reproduced with permission from [226]).

Microfluidics - the science of designing, manufacturing, and operating devices and processes that deal with small amounts of fluids (10^{-6} to 10^{-9} litres) - has the potential to significantly change the way of processing edible films. Microfluidic devices can be identified by the fact that they have channels with at least one dimension smaller than 1 mm. The devices themselves have dimensions ranging from millimeters down to micrometers. Recent investigations focused on generating and manipulating emulsions with microfluidic devices are motivated by the potential to use controlled flows and dispensing devices on the scale of the droplets or bubbles to tailor the properties of the emulsions and foams [224-225]. Examples of such a microfluidic device for creating a dispersion of water drops in oil or oil in water with starch granules are reported by Anna *et al.* [226] (shown in Figure 9) and Skurtys and Aguilera [227], respectively. The stream of the water is focused by the streams of oil into the flow-focusing orifice and breaks to release mono-dispersed or poly-dispersed water droplets into the outlet micro-channel. The oil forces the water to form a narrow thread that breaks off into mono-dispersed water droplets with diameters much smaller than the orifice diameter (Figure 9k and q). To prepare double emulsions, two T junctions can be used in a series as showed in Figure 10 [228].

Walther *et al.* [229] studied the deformation of droplets of a carrageenan solution immersed in sunflower oil (with different surfactant concentrations) after passing through a large flow-focusing orifice. Downstream the liquid carrageenan droplets were sheared by flow so that their shape was distorted.

Advantage of microfluidic techniques is that its possibility to produce a thin layer with a microstructures such as oil droplet, air bubbles (with or without aroma), encapsulated probiotics, etc. In the matrix, the distribution of the microstructures can be mono-dispersed in size and uniform dispersed therefore a homogeneous film is obtained. Moreover, the thickness of the layer can be also constant and controlled. Moreover, in comparison with the multilayer one process is necessary to produce the film.

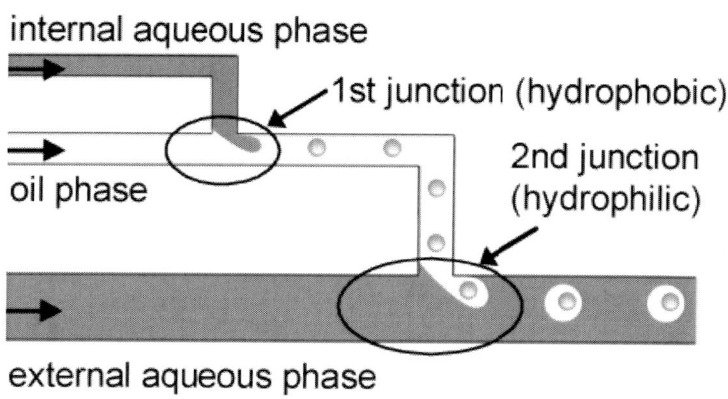

Figure 10. Scheme used to prepare double emulsions (W/O/W) using two T-shaped junctions in series (Reproduced with permission from [228]).

5.3. Control of Nanostructures for Edible Films

Solid-gel processing is widely used in the synthesis of inorganic and organic-inorganic hybrid materials and capable of producing nanoparticles, nanorods, thin films, and monolith. In food engineering and most particularly for the preparation of edible films, using atomic layer deposition can give to food scientists some advantages over conventional technologies. Atomic layer deposition (ALD) has demonstrated its capability of depositing 2 or more layers of materials and thus to create nanolaminates. Excellent reviews on ALD have been published by Ritala and Leskelä [230] for the materials science. As presented in the paragraph 4.3, the most interesting methods based on is based is the layer-by-layer (LbL) deposition technique which produce nanolaminates. The food surface is coated with interfacial films consisting of multiple nanolayers of different materials. The LbL deposition technology allows precise control over the thickness (0.2 to 500 nm), the composition, the structure and therefore over the properties of the interfacial films. Alloy macroscopic sample with a thickness up to 400 μm has been demonstrated.

5.4. Sensory Properties of Edible Coatings

Flavour is one of the most important characteristics of any food product. Its critical role in determining the way consumers assess food quality has made it a key area of research for the food industry [173]. However, the studies of sensory evaluation on the edible films and coatings are very scarce and more investigations need to be done. In particular, edible films can modified to flavour of the coated food and then affected the tolerability and the desirability by consumers. The taste, this is not the only sensory properties to investigate. There are also the colour and the texture of the coated food .

ACKNOWLEDGMENT

The authors (OS and CV) wish to thank CONICYT, sponsor of the Bicentenario Grant (PBCT PSD-62) and the project Innova Chile-CORFO CT11 PUT-20 .

REFERENCES

[1] Debeaufort, F.; Quezada-Gallo, J. A.; Voilley, A. *Crit. Rev. Food Sci.* 1998, *38*, 299–313.

[2] Tharanathan, R. N. *Trends Food Sci. Tech.* 2003, *14*, 71-78.

[3] Cha, D. S.; Chinnan, M. S. *Crit. Rev. Food Sci.* 2004, *44*, 223–237.

[4] Siracusa, V.; Rocculi, P.; Romani, S.; Dalla Rosa, M. *Trends Food Sci. Tech.* 2008, *19*, 634-643.

[5] Bifani, V.; Ramírez, C.; Ihl, M.; Rubilar, M.; Garcia, A.; Zaritzky, N. *LWT-Food Sci. Technol.* 2007, *40*, 1473-1481.

[6] Lin, D.; Zhao, Y. *Comp. Rev. Food Sci. Food Saf.* 2007, *6*, 60-75.

[7] Vargas, M.; Pastor, C.; Chiralt, A.; McClements, D. J.; González-Martínez, C. *Crit. Rev. Food Sci.* 2008, *48*, 496-511.

[8] Donhowe G; Fennema O. In *Edible coatings and films to improve food quality*; Krochta, J.M.; Baldwin, E.A.; Nisperos-Carriedo, M.O.; Ed.; Lancaster: Technomic Publishing, 1994; pp 1-24.

[9] Williams, P. A.; Phillips, G. O. In *Handbook of hydrocolloids*; Phillips G.; Willians P.; Ed.; CRC Press, Cambridge, England, 2000; pp 1-19.

[10] Kester, J. J.; Fennema, O. R. *Food Technol.* 1986, *40*, 47-59.

[11] Baldwin, E. A.; Nisperos-Carriedo, M. O.; Baker, R. A. *Crit. Rev. Food Sci.* 1995, 35, 509–524.

[12] Armisén, R.; Galatas, F. In *Handbook of Hydrocolloids*; Phillips G.; Willians P.; Ed.; CRC Press, Cambridge, England, 2000; pp 21-40.

[13] Stanley, N. F. In *Food polysaccharides and their applications*; Stephen A. M.; Ed.; New York, Marcel Dekker, 1995, pp 187-199.

[14] Freile-Pelegrín, Y.; Madera-Santana, T.; Robledo, D.; Veleva, L.; Quintana, P.; Azamar, J. A. *Polym. Degrad. Stabil.* 2007, *92*, 244-252.

[15] Geraldine, R. M.; Soares, N. F. F.; Botrel, D. A.; Goncalves, L. A. *Carbohyd. Polym.* 2008, *72*, 403-409.

[16] Phan The, D.; Debeaufort, F.; Luu, D.; Voilley, A. *J. Agr. Food Chem.* 2005, *53*, 973-981.

[17] Phan The, D.; Debeaufort, F.; Voilley, A.; Luu, D., *Food hydrocolloid* 2009, *23*, 691-699.

[18] King, A. H. In Food hydrocolloids; Glicksman M.; Ed.; CRC Press Inc., 1982, pp 115–88.

[19] Rhim, J. W. *LWT-Food Sci. Technol.* 2004, *37*, 323–30.

[20] Guilbert, S. *Bull. Int. Dairy Fed.* 2000, *346*, 10–16.

[21] Borchard, W.; Kenning, A.; Kapp, A.; Mayer, C. *Int. J. Biol. Macromol.* 2005, *35*, 247–56.

[22] Zactiti, E.; Kieckbusch, T. *J. Food Eng.* 2006, *77*, 462–467.

[23] Tapia, M. S.; Rojas-Graü, M. A.; Rodrígez, F. J.; Ramírez, J.; Carmona, A.; Martin-Belloso, O. *J. Food Sci.* 2007, *72*, 190-196.

[24] Pranoto, Y.; Salokhe, V. M.; Rakshit, S. K. *Food Res. Int.* 2005, *38*, 267–272.

[25] Rojas-Graü, M. A.; Avena-bustillos, R. J.; Olsen, C.; Friedman, M.; Henika, P.; Martín-Belloso, O.; Pan, Z.; McHugh, T. H. *J. Food Eng.* 2007, *81*, 634–641.

[26] Wu, Y.; Weller, C. L.; Hamouz, F.; Cuppett, S.; Schnepf, M. *J. Food Sci.* 2001, *66*, 486–93.

[27] Volery, P.; Besson, R.; Schaffer-Lequart, C. *J. Agr. Food Chem.* 2004, *52*, 7457-7463.

[28] Shaw, C.; Secrist, J.; Tuomy, J. *U.S. Patent 4196219*, 1980.

[29] Lee, J. Y.; Park, H. J.; Lee, C. Y.; Choi, W. Y. *LWT-Food Sci. Technol.* 2003, *36*, 323–329.

[30] Ninomiya, H.; Suzuki, S.; Ishii, K. *U.S. Patent 5620757*, 1997.

[31] Karbowiak, T.; Debeaufort, F.; Champion, D.; Voilley, A. *J. Colloid Interf. Sci.* 2006, *294*, 400–410.

[32] Karbowiak, T.; Debeaufort, F.; Voilley, A. *Food Hydrocolloid* 2007, *21*, 879–888.

[33] Ribeiro, C.; Vicente, A. A.; Teixeira, J. A.; Miranda, C. *Postharvest Biol. Tec.* 2007, *44*, 63–70.

[34] Murray J. C. F. In *Handbook of hydrocolloids*; Phillips G.; Willians P.; Ed.; CRC Press, Cambridge, England, 2000; pp 219-230.

[35] Gennadios, A.; Hanna, M. A.; Kurth, B. *LWT-Food Sci. Technol.* 1997, *30*, 337–50.

[36] Turhan, K. N.; Sahbaz, F. *J. Food Eng.* 2004, *61*, 459–466.
[37] Morillon, V.; Debeaufort, F.; Bond, G.; Capelle, M.; Voilley, A. *Crit. Rev. Food Sci.* 2002, *42*, 67–89.
[38] Debeaufort, F.; Quezada-Gallo, J. A.; Delporte, B.; Voilley, A. *J. Membrane Sci.* 2000, *180*, 47–55.
[39] Maftoonazad, N.; Ramaswamy, H. S. *LWT-Food Sci. Technol.* 2005, *38*, 617–24.
[40] Shahidi, F.; Arachchi, J. K. V.; Jeon, Y. J. *Trends Food Sci. Tech.* 1999, *10*, 37–51.
[41] Tharanathan, R. N.; Kittur, F. S. *Crit. Rev. Food Sci.* 2003, *43*, 61–87.
[42] Jayakumar, R.; Nwe, N. T.; Tokura, S.; Tamura, H. *Int. J. Biol. Macromol.* 2007, *40*, 175–181.
[43] Rabea, E. I.; Badawy, M. E. T.; Stevens, C. V.; Smagghe, G.; Steurbaut, W. *Biomacromolecules* 2003, *4*, 1457–1465.
[44] Tang, R.; Du, Y.; Fan, L. *J. Polym. Sci.* 2003, *41*, 993–997.
[45] Park, S. Y.; Marsh, K. S.; Rhim, J. W. *J. Food Sci.* 2002, *67*, 104–197.
[46] Peniston, Q. P.; Johnson, E. *US Patent 41951755*, 1980.
[47] Sathivel, S.; Liu, Q.; Huang, J.; Prinyawiwatkul, W. *J. Food Eng.* 2007, *83*, 366–373.
[48] Coma, V. *Meat Sci.* 2008, *78*, 90–103.
[49] Fan, W.; Sun, J.; Chen, Y.; Qiu, J.; Zhang, Y.; Chi, Y. *Food Chem.* 2009, *115*, 66-70.
[50] Fernandez-Saiz, P.; Lagaron, J. M.; Ocio, M. J. *Food hydrocolloid* 2009, *23*, 913-921.
[51] Coma, V.; Martial-Gros, A.; Garreau, S.; Copinet, A.; Deschamps, A. *J. Food Sci.* 2002, 67, 1162–1169.
[52] Sebti, I.; Martial-Gros, A.; Carnet-Pantiez, A.; Grelier, S.; Coma, V. *J. Food Sci.* 2006, *70*, 100–104.
[53] Bordenave, N.; Grelier, S.; Coma, V. *J. Agr. Food Chem.* 2007, *55*, 9479–9488.
[54] Han, C.; Zhao, Y.; Leonard, S. W.; Traber, M. G. *Postharvest Biol. Tec.* 2004, *33*, 67–78.
[55] Vargas, M.; Albors, A.; Chiralt, A.; González-Martinez, C. *Postharvest Biol. Tec.* 2006, *41*, 164–71.
[56] Chien, P. J.; Sheu, F.; Yang, F. H. *J. Food Eng.* 2007, *78*, 225–229.
[57] Dutta, P. K.; Tripathi, S.; Mehrotra, G. K.; Dutta, J. *Food Chem.* 2009, 114, 1173-1182.

[58] Subhashchandra, S. C.; Bhaskara, N.; Bhandary, M. H.; Raghunath, B. S. *J. Sci. Food Agr.* 1996, *70*, 453-460.

[59] Mei, Y.; Zhao, Y.; Yang, J.; Furr, H. R. *J. Food Sci.* 2002, *67*, 1964-1968.

[60] Sahin, S.; Summu, G.; Altunakar, B. *J. Sci. Food Agr.* 2005, *85*, 2375-2379.

[61] Bosquez-Molina, E.; Guerrero-Lagarreta, I.; Vernon-Carter, E. J. *Food Res. Int.* 2003, *36*, 885–893.

[62] Díaz-Sobac, R.; García, H. S.; Beristain, C.I.; Vernon-Carter, E. J. *J. Food Process. Pres.* 2002, *26*, 129–141.

[63] Mazza, G.; Qi, H. *J. Agr. Food Chem.* 1991, *39*, 2163–2166.

[64] Hoagland, P.; Parris, N. *J. Agr. Food Chem.* 1996, *44*, 1915–1919.

[65] Marudova, M.; Lang, S.; Brownsey, G. J.; Ring, S. G. *Carbohydr. Res.* 2005, *340*, 2144–2149.

[66] Rojas-Graü, M. A.; Avena-Bustillos, R. J.; Friedman, M.; Henika, P. R.; Martín-Belloso, O.; McHugh, T. H. *IFT Annual Meeting* 2006, Abstract *389.*

[67] Lenart, A.; Dabrowska, R. *Dry. Technol.* 1999, *7–8*, 1359–1373.

[68] Lenart, A.; Piotrowski, D. *Dry. Technol.* 2001, *19*, 849–877.

[69] Baeva, M.; Panchev, I. *Food Chem.* 2005, *92*, 343-348.

[70] Walstra, P. In *Physical Chemistry of Foods*; Walstra, P.; Ed.; Marcel Dekker - New York. 2003.

[71] Claudia, A. R. B.; Bello-Pérez, L. A.; Gacia, M. A.; Martino, M. N.; Solorza-Feria, J.; Zaritzky, N. E. *Carbohyd. Polym.* 2005, *60*, 235-244.

[72] Copeland, L.; Blazek, J.; Salman, H.; Tang, M. C. *Food Hydrocolloid* 2009, *23*, 1527-1534.

[73] Xu, X. Y.; Kim, K. M.; Hanna, M. A.; Nag, D. *Ind. Crop. Prod.* 2005, *21*, 185-192.

[74] Myllärinen, P.; Buléon, A.; Lahtinen, R.; Forssell, P. *Carbohyd. Polym.* 2002, *48*, 41–8.

[75] Rindlav, A.; Hulleman, S. H. D.; Gatenholm, P. *Carbohyd. Polym.* 1997, *34*, 25–30.

[76] Krogars, K.; Heinämäki, J.; Karjalainen, M.; Rantanen, J.; Luukkonen, P.; Yliruusi, J. *Eur J. Pharm. Biopharm.* 2003, *56*, 215–221.

[77] Parra, D. F.; Tadini, C. C.; Ponce, P.; Lagão, A. B. *Carbohyd. Polym.* 2004, *58*, 475–481.

[78] Mali, S.; Grossmann, M. V. E.; García, M. A.; Martino, M. N. M.; Zaritzky, N. E. *Food Hydrocolloid* 2005, *19*, 157-164.

[79] Hu, G.; Chen, J.; Gao, J. *Carbohyd. Polym.* 2009, *76*, 291–298.

[80] Baldwin, E. A.; Baker, R.A. In *Protein based films and coatings* Gennadios A.; Ed.; Boca Raton, Fla., CRC Press. 2002; pp 501–515.

[81] Sothornvit, R.; Krochta, J. M. *J. Agr. Food Chem.* 2000; *48*, 6298–302.

[82] Sothornvit, R.; Krochta, J. M. *J. Food Eng.* 2001, *50*, 149–155.

[83] Sothornvit, R.; Reid, D. S.; Krochta, J. M. *Amer. Soc. Ag. Eng.* 2002, *45*, 1479–1484.

[84] Cho, S. Y.; Rhee, C. *LWT-Food Sci. Technol.* 2002, *35*, 151–157.

[85] Wang, V. C. H.; Kim, M. S.; Lee, S. Y. *J. Food Sci.* 2005, *70*, 387–391.

[86] Gómez-Guillén, M. C.; Turnay, J.; Fernández-Martín, F.; Ulmo, N.; Lizarbe, M. A.; Montero, P. *Food Hydrocolloid* 2002, *16*, 25-34.

[87] Mitchell, J. R. In *Functional properties of food macromolecules* Hill, S. E.; Ledward, D. A.; Mitchell, J. R.; Ed.; Aspen Publishers., 1998; pp. 50–73.

[88] Sobral, P. J. A.; Habitante, A. M. Q. B. *Food Hydrocolloid* 2001, *15*, 377–382.

[89] Ledward, D. A., In *Handbook of hydrocolloids*; Phillips G.; Willians P.; Ed.; CRC Press, Cambridge, England, 2000; pp 67-86.

[90] Johnston-Banks, F.A., In *Gelatin* Harris, P.; Ed.; Elsevier, London, 1990; pp. 233–289.

[91] Sobral, P. J. A.; Menegalli, F. C.; Hubinger, M. D.; Roques, M. A. *Food Hydrocolloid* 2001, *15*, 423–432.

[92] Arvanitoyannis, I.S. In *Protein-based Films and Coatings* Gennadios, A.; Ed.; CRC Press, Boca Ratón, Florida, 2002.

[93] Simon-Lukasik, K. V.; Ludescher, R. D. *Food Hydrocolloid* 2004, *18*, 621–630.

[94] Gómez-Estaca, J.; Montero, P.; Fernández-Martín, F.; Gómez-Guillén, M. C. *J. Food Eng.* 2009, *90*, 480–486.

[95] Gómez-Guillén, M. C.; Ihl, M.; Bifani, V.; Silva, A.; Montero, P. *Food Hydrocolloid* 2007, *21*, 1133–1143.

[96] Pérez–Gago, M. B.; Nadaud, P.; Krochta, J. M. *J. Food Sci.* 1999, *64*, 1034–1037.

[97] Ennis, M. P.; Mulvihill, D. M. In *Handbook of hydrocolloids*; Phillips G.; Willians P.; Ed.; CRC Press, Cambridge, England, 2000; pp 189-218.

[98] Pérez-Gago, M. B.; Krochta, J. M. In *Protein-Based Films and Coatings* Gennadios, A.; Ed.; CRC Press, Boca Raton, FL, 2002; pp. 159–180.

[99] McHugh, T. H.; Krochta, J. M. *Food Technol.* 1994, *48*, 97–103.

[100] Fairley, P.; Monahan, F. J.; German, J. B.; Krochta, J. M. *J. Agr. Food Chem.* 1996, 44, 438–443.

[101] Miller, K. S.; Krochta, J. M. *J. Appl. Polym. Sci.* 1997, *88*, 1095–1099.

[102] Hong, S.-I., Krochta, J. M. *J. Food Sci.* 2003, *68*, 224–228.

[103] Shaw, N. B.; Monahan, F. J.; O'Riordan, E. D.; O'Sullivan, M. *J. Food Sci.* 2002, *67*, 164–167.

[104] Perez-Gago, M. B.; Krochta, J. M. *J. Food Sci.* 1999, *64*, 695–698.

[105] Ozdemir, M.; Floros, J. D. *J. Food Eng.* 2008, *84*, 116–123.

[106] Osés, J.; Fabregat-Vázquez, M.; Pedroza-Islas, R.; Tomás, S. A.; Cruz-Orea, A.; Maté, J. I. *J. Food Eng.* 2009, *92*, 56–62.

[107] Hagenmaier, R. D.; Shaw, P. E. *J. Agr. Food Chem.* 1991, 38, 1799–1803.

[108] Park, H. J. *Trends Food Sci. Tech.* 1999, *10*, 254–260.

[109] Cisneros-Zevallos, L.; Krochta, J. M. *J. Food Sci.* 2003, *68*, 503-510.

[110] Ofori-Kwakye, K.; Fell, J. T. *Int J Pharm* 2003, *250*, 251–257.

[111] Dogan, N.; McHugh, T.M. *J. Food Sci.* 2007, *72*, 16–22.

[112] Pinotti, A.; García, M. A.; Martino, M. N.; Zaritzky, N E. *Food Hydrocolloid* 2007, *21*, 66–72.

[113] Srinivasa, P. C.; Ramesh, M. N.; Tharanathan, R. N., *Food Hydrocolloid* 2007, *21*, 1113-1122.

[114] Faber, J. N.; Harris, L. J.; Parish, M. E.; Beuchat, L. R.; Suslow, T. V.; Gorney, J. R.; Garrett, E. H.; Busta, F.F. *Crit. Rev. Food Sci.* 2003, *2*, 142–160.

[115] Liu, L.; Kerry, J. K.; Kerry, J. P. *Int. J. Food Sci. Tech.* 2006, *41*, 295–302.

[116] Veiga-Santos, P.; Oliveira, L. M.; Cereda, M. P.; Alves, A. J.; Scamparini, A. R. P. *Food Hydrocolloid* 2005, *19*, 341–349.

[117] Chen, C. H.; Lai, L. S., *Food Hydrocolloid* 2008, *22*, 1584-1595.

[118] Carvalho, R. A.; Sobral, P. J. A.; Thomazine, M.; Habitante, A. M. Q. B.; Giménez, B. *Food Hydrocolloid* 2008, *22*, 1117-1123.

[119] Jongjareonrak, A.; Benjakul, S.; Visessanguan, W.; Prodpran, T.; Tanaka, M. *Food Hydrocolloid* 2006, *20*, 492–501.

[120] Barreto, P. L. M.; Pires, A. T. N.; Soldi, V. *Polym. Degrad. Stabil.* 2003, *79*, 147–152.

[121] Myllärinen, P.; Partanen, R.; Sepalla, J.; Forsell, P. *Carbohyd. Polym.* 2002, *50*, 355–61.

[122] Jagannath, J. H.; Nanjappa, C.; Das Gupta, D. K.; Bawa, A. S. *Int. J. Food Sci. Tech.* 2006, *41*, 498-506.

[123] Cerqueira, M. A.; Lima, A. M.; Souza, B. W. S.; Teixeira, J. A.; Moreira, R. A.; Vicente, A. A. *J. Agr. Food Chem.* 2009, *57*, 1456-1462.

[124] Bourtoom, T.; Chinnan, M. S.; Jantawat, P.; Sanguandeekul, R. *Food Sci. Technol. Int.* 2006, *12*, 119-126.

[125] Veiga-Santos, P.; Oliveira, L. M.; Cereda, M. P.; Scamparini, A. R. P. *Food Chem.* 2007, *103*, 255-262.

[126] Chang, Y. P.; Cheah, P.B.; Seow, C. C. *J. Food Sci.* 2000, *65*, 1–7.

[127] Krochta, J. M. In *Protein-based Films and Coatings* Gennadios A.; Ed.; Boca Raton, CRC Press, 2002; pp 1-41.

[128] Mali, S.; Grossmann, M. V. E.; García, M. A.; Martino, M. N. M.; Zaritzky, N. E. *J. Food Eng.* 2006, *75*, 453–60.

[129] Cha, D. S.; Chinnan, M. S. *Crit. Rev. Food Sci.* 2004, *44*, 223–237.

[130] Appendini, P.; Hotchkiss, J. H. *Innovative Food Science and Emerging Technologies* 2002, *3*, 113–126.

[131] Sivarooban, T.; Hettiarachchy, N. S.; Johnson, M. G. *Food Res. Int.* 2008, *41*, 781-785.

[132] Scannell, A. G. M.; Hill, C.; Ross, R. P.; Marz, S.; Hartmeier, W.; Arendt, E. K. *Int. J. Food Microbiol.* 2000, *60*, 241–249.

[133] Santiago-Silva, P.; Soares, N. F. F.; Nóbrega, J. E.; Júnior, M. A. W.; Barbosa, K. B. F.; Volp, A. C. P.; Zerdas, E. R. M. A.; Würlitzer, N. J. *Food Control* 2009, *20*, 85–89.

[134] Suppakul, P.; Miltz, J.; Sonneveld, K.; Bigger, S.W. *J. Food Sci.* 2003, *68*, 408–420.

[135] Mitrakas, G.E.; Koutsoumanis, K.P.; Lazarides, H.N. *Innovative Food Science and Emerging Technologies* 2008, *9*, 550–555.

[136] Mecitoğlu, C., Yemenicioğlu, A. *Food Chem.* 2007, *104*, 726–733.

[137] Seifu, E.; Buys, E. M.; Donkin, E. F. *Trends Food Sci. Tech.* 2005, *16*, 137-154.

[138] Kussendrager, K. D.; van Hooijdonk, A. C. M. *Brit J. Nutr.* 2000, *84*, 19–25.

[139] Seydim, A.C.; Sarikus, G. *Food Res. Int.* 2006, *39*, 639–644.

[140] Ayranci, E., Tunc, S. *Z Lebensm Unters Forsch A* 1997, *205*, 470-473.

[141] Bico, S. L. S.; Raposo, R. M. S.; Morais, C.; Morais, A. M. M. B. *Food Control* 2009, *20*, 508–514.

[142] Burt, S. A. *Int. J. Food Microbiol.* 2004, *94*, 223–53.

[143] Zinoviadou, K. G.; Koutsoumanis, K. P.; Biliaderis, C. G. *Meat Sci.* 2009, *82*, 338-345.

[144] Maizura, M.; Fazilah, A.; Norziah, M. H.; Karim, A. A. *J. Food Sci.* 2007, *72*, 324– 330.

[145] Ramesh, M. N.; Siddalingaiya, G. P. *Patent Wipo WO/2006/103698*; 2006.

[146] Park, S. I.; Zhao, Y. *J. Agr. Food Chem.* 2004, *52*, 1933–1939.

[147] Weaver, C. M.; Heaney, R. P. In *Modern Nutrition in Health and Disease* Shils M. E.; Olson J. A; Shike M.; Ross A. C.; Ed.; Philadelphia Lippincott Williams & Wilkins, 1998; pp 141-155.

[148] De Gennes, P. G.; Brochard-Wyart, F.; Quéré, D. In *Capillarity and Wetting Phenomena* De Gennes, P. G.; Brochard-Wyart, F.; Quéré, D.; Ed.; Springer-Verlag, 2004; pp 203.

[149] Miller, C. A.; Neogi, P. In *Interfacial phenomena: equilibrium and dynamic effects.* CRC Press Taylor & Francis Group, 2008; pp 61-107.

[150] Adamson, A. W.; Gast, A. P. In *Physical Chemistry of Surfaces*, Sixth Edition, Wiley-Interscience Publication, 1997; pp 347-430.

[151] Tghygesen, L. G.; Løkke, M. M.; Micklander, E.; Engelsen, S. B. *Trends Food Sci. Tech.* 2003, *14*, 50–57.

[152] McAnally, G. D.; Everall, N. J.; Chalmers, J. M.; Smith, W. E. *Appl. Spectrosc.* 2003, *57*, 44-49.

[153] Ishikawa, K.;Yamano, H.; Kagawa, K.; Asada, K.; Iwata, K.; Ueda, M. *Opt. Laser Eng.* 2004, *41*, 19–29.

[154] Hsu, B. L.; Weng, Y. M.; Liao, Y. H.; Chen, W. *J. Agr. Food Chem.* 2005, *53*, 5089–5095.

[155] Dai, H.; Shen, Y.; Zhou, H.; Cai, X. *Thin Solid Films* 2008, *516*, 1796–1802.

[156] Fukano, T.; Yamaguchi, I. *Opt. Lett.* 1966, *21*, 1942–1944.

[157] Kim, S. W.; Kim, G. H. *Appl. Optics* 1999, *38*, 5968–5973.

[158] Levich V. In *Physicochemical hydrodynamics*. Prentice-Hall, 1962.

[159] Snoeijer, J. H.; Ziegler, J.; Andreotti, B.; Fermigier, M.; Eggers, J. *Phys. Rev. Lett.* 2008, *100*, 244502.

[160] Sperling, L.H. In *Introduction to Physical Polymer Science*. 2nd ed., New York, NY, USA: John Wiley & Sons, Inc, 2006; Chp 11.

[161] Kristo, E.; Biliaderis, C. G.; Zampraka, A. *Food Chem.* 2007, *101*, 753–76.

[162] Butler, B. L.; Vergano, P. J.; Testin, R. F.; Bunn, J. M.; Wiles, J. L. *J. Food Sci.* 1996, *61*, 953-61.

[163] Park J. W.; Testin R. F.; Park H. J.; Vergano P. J.; Weller C. L. *J. Food Sci.* 1994, *59*, 916–919.

[164] Mark, J. E. In *Polymer Data Handbook.* Oxford University Press, Inc, 1999.

[165] Lazaridou, A.; Biliaderis, C. G. *Carbohydr. Polym.* 2002, *48*, 179-190.

[166] Wang, Y.; Padua, G. W. *Macromol. Mater. Eng.* 2003, *288*, 886–893.

[167] Pfeiffer, T.; Menner, M. *Fleischwirtschaft* 1999, *79*, 79-84.

[168] Crank, J. In *The Mathematics of Diffusion.* 2nd ed; Clarendon Press, Oxford, 1975.

[169] Lide, D. R. In *Handbook of Chemistry and Physics.* 87th Edition, CRC Taylor and Francis, 2007.

[170] Talasila, P. C.; Cameron, A. C. *J. Food Sci.* 1997, *62*, 926-930.

[171] Simpson, R.; Acevedo, C.; Almonacid, S. *J. Food Eng.* 2009, *92*, 233-239.

[172] Gontard, N.; Guilbert, S.; Cuq, J. L. *J. Food Sci.* 1993, *58*, 206–211.

[173] Voilley, A.; Souchon, I. In *Flavour in Food.* Voilley, A.; Etiévant, P.; Ed.; Woodhead Publishing Limited and CRC Press LLC, 2006; pp 117-132.

[174] Hirose, K.; Harte, B. R.; Giacin, J. R.; Miltz, J.; Stine, C. In *Food and Packaging Interactions*, Hotchkiss J. H.; Ed.; ACS Symposium Series 365, American Chemical Society, Washington, DC, 1998; pp 28-41.

[175] Miller, K. S.; Upadhyaya, S. K.; Krochta, J. M. *J. Food Sci.* 1998, *63*, 244-247.

[176] Williams, R.; Mittal, G. S. *LWT-Food Sci. Technol.* 1999, *32*, 440–445.

[177] Han, C.; Krochta, J. M. *J. Food Sci.* 2001, *66*, 294-299.

[178] García, M. A.; Ferrero, C.; Bértola, N.; Martino, M.; Zaritzky, N. *Innovative Food Science and Emerging Technologies* 2002, *3*, 391–397.

[179] García, M.; Bifani, V.; Campos, C.; Martino, M. N.; Sobral, P.; Flores, S.; Ferrero, C.; Bertola, N.; Zaritzky, N. E.; Gerschenson, L.; Ramírez, C.; Silva, A.; Ihl, M.; Menegalli, F. In *Food Engineering: Integrated Approches*, Gutiérrez-López, G. F.; Barbosa-Cánovas, G. V.; Welti-Chanes, J.; Parada-Arias, E.; Ed.; Springer, 2008; pp 225-241.

[180] Trezza, T. A.; Krochta, J. M. *J. Food Sci.* 2000, *65*, 658-662.

[181] Nikolova, K.; Panchev, I.; Sainov, S. *J. Optoelectron. Adv. M.* 2005, *7*, 1439-1444.

[182] Villalobos, R.; Chanona, J.; Hernández, P.; Gutiérrez, G.; Chiralt, A. *Food Hydrocolloid* 2005, *19*, 53–61.

[183] Fabra, M. J.; Talens, P.; Chiralt, A. *Food Hydrocolloid* 2009, *23*, 676–683.

[184] Hutchings, J. B. In *Food color and appearance,* Maryland: Aspen Publisher, 1999.

[185] Enrione, J. I. In PhD thesis *Mechanical Stability of Intermediate Moisture Starch-Glycerol Systems.* Division of Food Sciences. Nottingham, The University of Nottingham, 2005.

[186] Anderson, R. B. *J. Am. Chem. Soc.* 1946, *68*, 686-691.

[187] De Boar, J. H. In *The Dynamic Character of Adsorption*, Oxford, UK, Clarendon Press, 1953.

[188] Filho, J. F. L.; Romanelli, P. F.; Barboza, S. H. R.; Gabas, A. L.; Telis-Romero, J. *J. Food Eng.* 2002, *52*, 201-206.

[189] Vazquez, G.; Chenlo, F.; Moreira R. *J. Food Eng.* 2003, *60*, 449-452.

[190] Quirijns, E.; van Boxtel, A.; van Loon, W.; van Straten, G. *J. Sci. Food Agr.* 2005, *85*, 1805-1814.

[191] Enrione, J. I.; Hill, S. E.; Mitchell, J. R. *J. Agr. Food Chem.* 2007, *55*, 2956-2963.

[192] Yan, Z.; Soussa-Gallagher, J.; Oliveira Fernanda, A. R. *J. Food Eng.* 2008, *86*, 342-348.

[193] Gabas, A. L.; Menegalli, F. C.; Telis-Romero, J. *J. Food Sci.* 2000*, 65*, 680-684.

[194] Tunc, S.; Duman, O. *J. Food Eng.* 2007, *81*, 133-143.

[195] McMinn, W. A. M.; Al-Muhtaseb, A. H.; Magee, T. R. A. *Food Res. Int.* 2005, *38*, 505-510.

[196] Nkolo, M. E. Y. N.; Noah, N. J.; Bardet, S. *Thermochim. Acta* 2008, *468*, 1-5.

[197] Liu, Y.; Feng. *J. Mater. Sci.* 2001, *36*, 1809-1815.

[198] Van den Berg, C.; Bruin, S. In *Water Activity and its Estimation in Food Systems:Theoretical Aspects,* Rockland, L. B.; Stewart, G. F.; Ed.; London, UK, Academic Press, 1981.

[199] Krug, R. R.; Hunter, W. G.; Greiger, R. A. *J. Phys. Chem.* 1976, *80*, 2335-2341.

[200] Krug, R. R.; Hunter, W. G.; Greiger, R. A. *J. Phys. Chem.* 1976*, 80*, 2341-2352.

[201] Leffler, J. *J. Org. Chem.* 1955, *20*, 1202.

[202] Cha, D. S.; Cooksey, K.; Chinnan, M. S.; Park, H. J. *LWT-Food Sci. Technol.* 2003, *36*, 209–213.

[203] Tharanathan, R. N.; Srinivasa, P. C.; Ramesh, M. N. *Indian patent 0085/DEL/02*; 2002.

[204] Mastromatteo, M.; Chillo, S.; Buonocore, G. G.; Massaro, A.; Conte, A.; Bevilacqua, A.; Del Nobile, M. A. *J. Food Eng.* 2009, *92*, 467–473.

[205] Obara, S.; McGinity, J. W. *Int. J. Pharm.* 1995, *37*, 849-53.

[206] Alcantara, C. R.; Rumsey, T. R.; Krochta, J. M. *J. Food Process. Engr.* 1998, *21*, 387-405.

[207] Vargas, M.; Albors, A.; Chiralt, A.; González-Martinez, C. *Food Hydrocolloid* 2009, *23*, 536–547.

[208] Chambi, H.; Grosso, C. *Food Res. Int.* 2006, *39*, 458–466.

[209] Kaya, S.; Kaya, A. *J. Food Eng.* 2000, *43*, 91-96.

[210] Finkenstadt, V. L.; Willett, J. L. *Carbohydr. Polym.* 2004, *55*, 149–154.

[211] Guan, J.; Hanna, M. A. *Bioresource Technol.* 2006, *97*, 1716–1726.

[212] Fishman, M. L.; Coffin, D. R.; Onwulata, C. I.; Konstance, R. P. *Carbohydr. Polym.* 2004, *57*, 401–413.

[213] Naga, M.; Kirihara, S.; Tokugawa, Y.; Tsuda, F.; Hirotsuka., M. *US Patent 5569482*; 1996.

[214] Koepff, P.; Braumer, K.; Stahl, H., Dick, E. *US Patent 5316717*; 1994.

[215] Prakash, O. Comput. Mat. Sci. 2006, *37*, 7–11.

[216] Ouattara, B.; Simard, R. E.; Piette, G.; Bégin, A.; Holley, R. A. *Int. J. Food Microbiol.* 2000, *62*, 139-48.

[217] Weiss, J.; Takhistov, P.; McClements, D. J. *J. Food Sci.* 2006, *71*, 107–116.

[218] Stone, H. A. *Annu. Rev. Fluid Mech.* 1994, *26*, 65–102.

[219] Bergeron, V.; Bonn, D.; Martin, J. Y.; Vovelle, L. *Nature* 2000, *405*, 772.

[220] Bergeron, V. *C. R. Physique*, 2003, *4*, 211–219.

[221] Romanazzi, G.; Nigro, F.; Ippolito, A.; Divenere, D.; Salerno, M. *J. Food Sci.* 2002, *67*, 1862-1867.

[222] Meng, X.; Li, B.; Liu, J.; Tian, S. *Food Chem.* 2008, *106*, 501–508.

[223] Bhaskara Reddy, M. V.; Belkacemi, K.; Corcuff, R.; Castaigne, F.; Arul, J. *Postharvest Biol. Tec.* 2000, *20*, 39–51.

[224] Skurtys, O.; Aguilera, J. M. *J. Colloid Interf. Sci.* 2008. *318*, 380.

[225] Skurtys, O.; Aguilera, J. M. *Food Biophysics* 2008, *3*, 1.

[226] Anna, S. L.; Bontoux, N.; Stone, H. A. *Appl. Phys. Lett.* 2003, *82*, 364 – 366.

[227] Skurtys, O.; Aguilera, J. M. *Food hydroccoloid* 2009, In Press, doi:10.1016/j.foodhyd.2009.01.014

[228] Okushima, S.; Nisisako, T.; Torii T.; Higuchi, T. *Langmuir* 2004, *20*, 9905–9908.

[229] Walther, B. L.; Walkenström, P.; Hermansson, A. M. *J. Colloid Interf. Sci.* 2004, *270*, 195 – 204.

[230] Ritala, M.; Leskelä, M. In *Handbook of Thin Film Materials*, Nalwa, H. S.; Ed; Academic Press, San Diego, CA, Vol 1, 2002; pp.103.

INDEX

A

B

H

I

K

L

M

N

O

P

Q

R

S

T

U

V

W

X

Z